Computational simulation of scientific phenomena and engineering problems often depend on solving linear systems with a large number of unknowns. This book gives an insight into the construction of iterative methods for the solution of such systems and helps the reader to select the best solver for given classes of problems.

The emphasis is on the main ideas and how they have led to efficient solvers such as CG, GMRES, and Bi-CGSTAB. The author also explains the main concepts behind the construction of preconditioners. The reader is encouraged to build his own experience by analysing numerous examples that illustrate how best to exploit the methods. The book also hints at many open problems and, as such, it will appeal to established researchers. There are many exercises that clarify the material and help students to understand the essential steps in the analysis and construction of algorithms.

CAMBRIDGE MONOGRAPHS ON
APPLIED AND COMPUTATIONAL
MATHEMATICS

13 Iterative Krylov Methods for Large Linear Systems

The *Cambridge Monographs on Applied and Computational Mathematics* series reflects the crucial role of mathematical and computational techniques in contemporary science. The series publishes expositions on all aspects of applicable and numerical mathematics, with an emphasis on new developments in this fast-moving area of research.

State-of-the-art methods and algorithms as well as modern mathematical descriptions of physical and mechanical ideas are presented in a manner suited to graduate research students and professionals alike. Sound pedagogical presentation is a prerequisite. It is intended that books in the series will serve to inform a new generation of researchers.

Also in this series:

1. A practical guide to pseudospectral methods, *Bengt Fornberg*
2. Dynamical systems and numerical analysis, *A. M. Stuart & A. R. Humphries*
3. Level set methods and fast marching methods (2nd Edition), *J. A. Sethian*
4. The numerical solution of integral equations of the second kind, *Kendall E. Atkinson*
5. Orthogonal rational functions, *Adhemar Bultheel et al.*
6. The theory of composites, *Graeme W. Milton*
7. Geometry and topology for mesh generation, *Herbert Edelsbrunner*
8. Schwarz–Christoffel mapping, *Tobin A. Driscoll & Lloyd N. Trefethen*
9. High-order methods for incompressible fluid flow, *M. O. Deville, P. F. Fischer & E. H. Mund*
10. Practical extrapolation methods, *Avram Sidi*
11. Generalized Riemann problems in computational fluid dynamics, *Matania Ben-Artzi & Joseph Falcovitz*
12. Radial basis functions, *Martin D. Buhmann*
13. Iterative Krylov methods for large linear systems, *Henk van der Vorst*
14. Simulating Hamiltonian dynamics, *Benedict Leimkuhler & Sebastian Reich*
15. Collocation methods for Volterra integral and related functional differential equations, *Hermann Brunner*
16. Topology for computing, *Afra J. Zomorodian*
17. Scattered data approximation, *Holger Wendland*
19. Matrix preconditioning techniques and applications, *Ke Chen*
21. Spectral methods for time-dependent problems, *Jan Hesthaven, Sigal Gottlieb & David Gottlieb*
22. The mathematical foundations of mixing, *Rob Sturman, Julio M. Ottino & Stephen Wiggins*
23. Curve and surface reconstruction, *Tamal K. Dey*
24. Learning theory, *Felipe Cucker & Ding Xuan Zhou*
25. Algebraic geometry and statistical learning theory, *Sumio Watanabe*

Iterative Krylov Methods for Large Linear Systems

Henk A. van der Vorst
Utrecht University

Shaftesbury Road, Cambridge CB2 8EA, United Kingdom

One Liberty Plaza, 20th Floor, New York, NY 10006, USA

477 Williamstown Road, Port Melbourne, VIC 3207, Australia

314–321, 3rd Floor, Plot 3, Splendor Forum, Jasola District Centre, New Delhi – 110025, India

103 Penang Road, #05–06/07, Visioncrest Commercial, Singapore 238467

Cambridge University Press is part of Cambridge University Press & Assessment, a department of the University of Cambridge.

We share the University's mission to contribute to society through the pursuit of education, learning and research at the highest international levels of excellence.

www.cambridge.org
Information on this title: www.cambridge.org/9780521183703

First published 2003
Paperback edition 2009

A catalogue record for this publication is available from the British Library

Library of Congress Cataloging-in-Publication data
Vorst, H. A. van der, 1944–
Iterative Krylov methods for large linear systems / Henk A. van der Vorst.
p. cm. – (Cambridge monographs on applied and computational mathematics; v. 13)
ISBN 978 0 521 81828 5
1. Iterative methods (Mathematics) 2. Linear systems. I. Title.
II. Cambridge monographs on applied and computational mathematics; 13.
QA297.8. V67 2003
511′.4–dc21 2002035013

ISBN 978-0-521-81828-5 Hardback
ISBN 978-0-521-18370-3 Paperback

Contents

Preface

In 1991 I was invited by Philippe Toint to give a presentation, on Conjugate Gradients and related iterative methods, at the university of Namur (Belgium). I had prepared a few hand-written notes to guide myself through an old-fashioned presentation with blackboard and chalk. Some listeners asked for a copy of the notes and afterwards I heard from Philippe that they had been quite instructive for his students. This motivated me to work them out in LaTeX and that led to the first seven or so pages of my lecture notes. I took the habit of expanding them before and after new lectures and after I had read new interesting aspects of iterative methods. Around 1995 I put the then about thirty pages on my website. They turned out to be quite popular and I received many suggestions for improvement and expansion, most of them by e-mail from various people: novices in the area, students, experts in this field, and users from other fields and industry.

For instance, research groups at Philips Eindhoven used the text for their understanding of iterative methods and they sometimes asked me to comment on certain novel ideas that they had heard of at conferences or picked up from literature. This led, amongst others, to sections on GPBi-CG, and symmetric complex systems. Discussions with colleagues about new developments inspired me to comment on these in my Lecture Notes and so I wrote sections on Simple GMRES and on the superlinear convergence of Conjugate Gradients.

A couple of years ago, I started to use these Lecture Notes as material for undergraduate teaching in Utrecht and I found it helpful to include some exercises in the text. Eventually, the text grew larger and larger and it resulted in this book.

The history of the text explains more or less what makes this book different from various other books. It contains, of course, the basic material and the required theory. The mathematical presentation is very lecture inspired in the sense that I seldom prove theorems in lectures: I rather present the successive ideas in a way that appears logical, at least to me. My presentation of the Bi-CG method is an example of this. Bi-CG can be presented in a very short and clean way once the method is known. However, I preferred to introduce the method from the point of view of the person who only knows CG and has learned that CG does not work for unsymmetric systems. The natural question then is how to repair the method, retaining its short recurrences, so that unsymmetric problems can be solved. This is exemplary for the major part of the text. In most cases I try to present the ideas in the way they may come up in a discovery phase and then I, sometimes, collect interesting conclusions in theorems. I only included in some cases a more formal ('afterward') proof.

The text contains the basic material for the best known iterative methods and most of them are shown in frames in a way that facilitates implementation in Matlab. The experiments have been carried out with my own Matlab versions of the framed algorithms. The experiments are also rather unusual. They are very simple and easy to repeat. They are of a kind that anyone can come up with rather quickly. However, by inspecting the behaviour of the various methods for these simple problems, we observe various aspects that are typical for the more complicated real-life problems as well. I have seen many large and complicated linear systems and I have advised researchers in the industry on how to solve these systems. I found it always instructive to explain expected effects with the help of very simple small examples. I expect that these experiments and the discussions will add more life to the material for industrial users and that they will help students to construct other, even more interesting, test examples. I hope too that my discussions will stimulate students to discover other new aspects and to think about these.

Apart from the basic material, the text focuses on aspects that I found particularly interesting. Mostly these are aspects that lead to more insight or to better methods, but sometimes I have also included discussions on ideas (of others), which have less certain outcomes. Hence the book may also be of interest to researchers, because it hints at many avenues for new research.

I know that some of my colleagues have used older versions of this text for teaching, at various levels. For that purpose I have included all sorts of exercises in the text. As a student I did not like many of the exercises that we had to do, which served as tests afterwards. I preferred exercises that helped me understand the text when I needed it most. With this in mind I have constructed the exercises. In most cases they concern essential parts of the presentation and they are placed in the text where their results are most helpful for deeper understanding. Often I refer to results and formulas obtained in these exercises. Other exercises are intended to motivate students to construct working examples and to teach them to draw, and sometimes prove, correct conclusions. These exercises can be made with the help of short Matlab, or Mathematica, codes that the students have to write themselves, guided by the framed algorithms. Some of the problems can also be handled with the existing Template codes for various Krylov subspace methods, including CG, Bi-CG, GMRES, CGS, QMR, and Bi-CGSTAB. These methods are standard and are available in Matlab 6.0 and more recent versions. They are also available through the Web at the famous netlib website.

This is the place to thank numerous colleagues for their suggestions and comments. Over the past ten years they have been so numerous that it is impossible to mention all of them and it would also be unfair not to mention the many persons from various audiences that helped me, through their questions and comments, to improve my presentation of the material. I would like to make a few exceptions. I am particularly indebted to Michael Saunders, who helped me to correct part of the text while we were at a workshop in Copenhagen. I also learned very much about the subject from collaborations with Iain Duff, Gene Golub, and Youcef Saad, with whom I wrote a number of overview papers. These were extremely pleasant and fruitful events.

Dear reader of this book, I hope very much that this text will be helpful to you and I would appreciate hearing your comments and suggestions for further improvement.

Henk van der Vorst

1
Introduction

In this book I present an overview of a number of related iterative methods for the solution of linear systems of equations. These methods are so-called Krylov projection type methods and they include popular methods such as Conjugate Gradients, MINRES, SYMMLQ, Bi-Conjugate Gradients, QMR, Bi-CGSTAB, CGS, LSQR, and GMRES. I will show how these methods can be derived from simple basic iteration formulas and how they are related. My focus is on the ideas behind the derivation of these methods, rather than on a complete presentation of various aspects and theoretical properties.

In the text there are a large number of references for more detailed information. Iterative methods form a rich and lively area of research and it is not surprising that this has already led to a number of books. The first book devoted entirely to the subject was published by Varga [212], it contains much of the theory that is still relevant, but it does not deal with the Krylov subspace methods (which were not yet popular at the time).

Other books that should be mentioned in the context of Krylov subspace methods are the 'Templates' book [20] and Greenbaum's book [101]. The Templates are a good source of information on the algorithmic aspects of the iterative methods and Greenbaum's text can be seen as the theoretical background for the Templates.

Axelsson [10] published a book that gave much attention to preconditioning aspects, in particular all sorts of variants of (block and modified) incomplete decompositions. The book by Saad [168] is also a good source of information on preconditioners, with much inside experience for such methods as threshold ILU. Of course, GMRES receives much attention in [168], together with variants of the method. Kelley [126] considers a few of the most popular Krylov methods and discusses how to use them for nonlinear systems. Meurant [144] covers the theory of most of the best algorithms so far known. It contains extensive material on domain decomposition and multilevel type preconditioners. Meurant's book is also very useful as a source text: it contains as many as 1368 references to literature. Brezinski's book [31] emphasizes the relation between (Krylov) subspace methods and extrapolation methods. He also considers various forms of hybrid methods and discusses different approaches for nonlinear systems. Implementation aspects for modern High-Performance computers are discussed in detail in [61].

For general background on linear algebra for numerical applications see [98, 181], and for the effects of finite precision, for general linear systems, I refer to [116] (as a modern successor of Wilkinson's book [222]).

Some useful state of the art papers have appeared; I mention papers on the history of iterative methods by Golub and van der Vorst [97], and Saad and van der Vorst [170]. An overview on parallelizable aspects of sparse matrix techniques is presented in [70]. A state-of-the-art overview for preconditioners is presented in [22].

Iterative methods are often used in combination with so-called preconditioning operators (easily invertible approximations for the operator of the system to be solved). I will give a brief overview of the various preconditioners that exist.

The purpose of this book is to make the reader familiar with the ideas and the usage of iterative methods. I expect that then a correct choice of method can be made for a particular class of problems. The book will also provide guidance on how to tune these methods, particularly for the selection or construction of effective preconditioners.

For the application of iterative schemes we usually have linear sparse systems in mind, for instance linear systems arising in the finite element or finite difference approximations of (systems of) partial differential equations. However, the structure of the operators plays no explicit role in any of these schemes, which may also be used successfully to solve certain large dense linear systems. Depending on the situation, this may be attractive in terms of numbers of floating point operations.

I will also pay some attention to the implementation aspects of these methods, especially for parallel computers.

Before I start the actual discussion of iterative methods, I will first give a motivation for their use. As we will see, iterative methods are not only great fun to play with and interesting objects for analysis, but they are really useful in many situations. For truly large problems they may sometimes offer the only way towards a solution, as we will see.

Figure 1.1. The computational grid for an ocean flow.

1.1 On the origin of iterative methods

In scientific computing most computational time is spent in solving systems of linear equations. These systems can be quite large, for instance as in computational fluid flow problems, where each equation describes how the value of a local unknown parameter (for example the local velocity of the flow) depends on (unknown) values in the near neighbourhood.

The actual computation is restricted to values on a previously constructed grid and the number of gridpoints determines the dimensions of the linear system. In Figure. 1.1 we see such a grid for the computation of two-dimensional ocean flows. Each gridpoint is associated with one or more unknowns and with equations that describe how these unknowns are related to unknowns for neighbouring gridpoints. These relations are dictated by the physical model. Because many gridpoints are necessary in order to have a realistic computational model, we will as a consequence have many equations. A nice property of these linear systems is that each equation contains only a few unknowns. The matrix of the system contains mainly zeros. This property will be of great importance for the efficiency of solution methods, as we will see later.

We see that the grid consists of four differently represented subgrids. The reason for this is that, in the actual computations for this problem, we had to do the work in parallel: in this case on four parallel computers. This made it possible to do the work in an acceptably short time, which is convenient for model studies. We will see that most of the methods that we will describe lend themselves to parallel computation.

As we will see, the process of solving the unknowns from these large linear systems involves much computational work. The obvious approach via direct Gaussian elimination is often not attractive. This was already recognized by the great Gauss himself, in 1823, albeit for different reasons to those in the present circumstances [93]. In that year he proposed an iterative method for the solution of four equations with four unknowns, arising from triangular measurements.

In order to appreciate his way of computing, we start with the familiar (Gaussian) elimination process. As an example we consider the small linear system

$$\begin{bmatrix} 10 & 0 & 1 \\ \frac{1}{2} & 7 & 1 \\ 1 & 0 & 6 \end{bmatrix} \begin{bmatrix} x_1 \\ x_2 \\ x_3 \end{bmatrix} = \begin{bmatrix} 21 \\ 9 \\ 8 \end{bmatrix}.$$

The elimination process is as follows. We subtract $\frac{1}{20}$ times the first row from the second row and then $\frac{1}{10}$ times the first row from the third row. After this we have zeros in the first column below the diagonal and the system becomes

$$\begin{bmatrix} 10 & 0 & 1 \\ 0 & 7 & 1 - \frac{1}{20} \\ 0 & 0 & 6 - \frac{1}{10} \end{bmatrix} \begin{bmatrix} x_1 \\ x_2 \\ x_3 \end{bmatrix} = \begin{bmatrix} 21 \\ 9 - \frac{21}{20} \\ 8 - \frac{21}{10} \end{bmatrix}.$$

As a coincidence we also have a zero element in the second column below the diagonal, and now we can solve the system without much effort. It leads to the solution $x_3 = 1, x_2 = 1$, and $x_1 = 2$. Note that we have used exact computation. This is not a problem in this case, which has been designed to have a 'nice' solution. However, in more realistic situations, we may have non-integer values and then exact computation may lead to significant computational effort for a human being. It is not so easy to avoid human errors and after checking that the computed erroneous solution does not satisfy the initial system, it is not easy to find the place where the error occurred. Gauss suffered from this in his computations. He had a good physical intuition and he knew that the solution of his system should have components of about the same order of magnitude. Because his matrices had strongly dominating elements on the diagonal, he knew that the main contribution in the right-hand side came from the components in the solution that had been multiplied by a diagonal element: in our example $10x_1$, $7x_2$, and $6x_3$. This implies that if we neglect the off-diagonal elements in our system,

$$\begin{bmatrix} 10 & 0 & 0 \\ 0 & 7 & 0 \\ 0 & 0 & 6 \end{bmatrix} \begin{bmatrix} x_1 \\ x_2 \\ x_3 \end{bmatrix} = \begin{bmatrix} 21 \\ 9 \\ 8 \end{bmatrix},$$

we may still expect, from this perturbed system, a fairly good approximation for the solution of the unperturbed system; in our example: $x_1 = 2.1$, $x_2 = \frac{9}{7}$, and $x_3 = \frac{8}{6}$. Indeed, this is a crude approximation for the solution that we want. This way of approximation is still popular; it is known as a Gauss–Jacobi approximation, because the mathematician-astronomer Jacobi used it for the computation of perturbations in the orbits of the planets in our solar system.

Gauss made another intelligent improvement. He observed that we can approximate the original system better if we only replace nonzero elements in the strict upper triangular part. This leads to

$$\begin{bmatrix} 10 & 0 & 0 \\ \frac{1}{2} & 7 & 0 \\ 1 & 0 & 6 \end{bmatrix} \begin{bmatrix} x_1 \\ x_2 \\ x_3 \end{bmatrix} = \begin{bmatrix} 21 \\ 9 \\ 8 \end{bmatrix}.$$

This system has the solution $x_1 = 2.1$, $x_2 = \frac{7.95}{7}$, and $x_3 = \frac{5.9}{6}$. Indeed, this leads to an improvement (it should be noted that this is not always the case; there are situations where this approach does not lead to an improvement). The approach is known as the Gauss–Seidel approximation.

Altogether, we have obtained a crude approximated solution for our small system for only a small reduction in the computational costs. At this point it is good to discuss the computational complexity. For a system with n equations and n unknowns we need $2(n-1)^2$ operations to create zeros in the first column (if we ignore possible, already present, zeros). Then for the second column we need $2(n-2)^2$ operations. From this we conclude that for the elimination of all elements in the lower triangular part, we need about $\frac{2}{3}n^3$ operations. The cost of solving the resulting upper triangular system again requires roughly n^2 operations, which is a relatively minor cost for larger values of n. We may conclude that the cost of obtaining the exact solution is proportional to n^3. It is easy to see that the cost of computing only the Gauss–Seidel approximation is proportional to n^2 and it may be seen that this promises great advantages for larger systems.

The question now arises – how is the obtained approximated solution improved at relatively low costs? Of course, Gauss had also considered this aspect. In order to explain his approach, I will use matrix notation. This was not yet invented in Gauss's time and the lack of it makes the reading of his original description not so easy. We will write the system as

$$Ax = b,$$

with

$$A = \begin{bmatrix} 10 & 0 & 1 \\ \frac{1}{2} & 7 & 1 \\ 1 & 0 & 6 \end{bmatrix},$$

$$x = \begin{bmatrix} x_1 \\ x_2 \\ x_3 \end{bmatrix} \quad \text{and} \quad \begin{bmatrix} 21 \\ 9 \\ 8 \end{bmatrix}.$$

The lower triangular part of A is denoted by L:

$$L = \begin{bmatrix} 10 & 0 & 0 \\ \frac{1}{2} & 7 & 0 \\ 1 & 0 & 6 \end{bmatrix}.$$

The Gauss–Seidel approximation is then obtained by solving the system

$$L\widetilde{x} = b.$$

For a correction to this solution we look for the 'missing' part Δx:

$$A(\widetilde{x} + \Delta x) = b,$$

and this missing part satisfies the equation

$$A\Delta x = b - A\widetilde{x} \equiv r.$$

It is now an obvious idea to compute an approximation for Δx again with a Gauss–Seidel approximation, that is we compute $\widetilde{\Delta x}$ from

$$L\widetilde{\Delta x} = r,$$

and we correct our first approximation with this approximated correction $\widetilde{x}$:

$$\widetilde{\widetilde{x}} = \widetilde{x} + \widetilde{\Delta x}.$$

Table 1.1. Results for three Gauss–Seidel iterations

iteration	1	2	3
x_1	2.1000	2.0017	2.000028
x_2	1.1357	1.0023	1.000038
x_3	0.9833	0.9997	0.999995

Of course, we can repeat this trick and that leads to the following simple iteration procedure:

$$x^{(i+1)} = x^{(i)} + L^{-1}(b - Ax^{(i)}),$$

where the vector $y = L^{-1}(b - Ax^{(i)})$ is computed by solving

$$Ly = b - Ax^{(i)}.$$

We try this process for our little linear system. In the absence of further information on the solution, we start with $x^{(0)} = 0$. In Table 1.1 we display the results for the first three iteration steps.

We observe that in this case we improve the solution by about two decimals per iteration. Of course, this is not always the case. It depends on how strongly the diagonal elements dominate. For instance, for the ocean flow problems we have almost no diagonal dominance and Gauss–Seidel iteration is so slow that it is not practical in this bare form.

The computational costs per iteration step amount to roughly $2n^2$ operations (additions, subtractions, multiplications) for the computation of $Ax^{(i)}$, plus n^2 operations for the solution of the lower triangular system with L: in total $\approx 3n^2$ operations per iteration step. Solution via the direct Gaussian elimination process takes $\approx \frac{2}{3}n^3$ operations. This implies a gain in efficiency if we are satisfied with the approximations and if these are obtained after less than

$$\left(\frac{2}{3}n^3\right)/(3n^2) = \frac{2}{9}n$$

iterations.

Computation with this iteration method was very attractive for Gauss, not because of efficiency reasons but mainly because he did not have to compute the approximated solutions accurately. A few decimal places were sufficient. Unintentional errors and rounding errors 'correct themselves' in the later iterations. Another advantage is that the residual $b - Ax^{(i)}$ has to be computed in each step, so that we can see at a glance how well the computed approximation satisfies the system. In a letter to his colleague Gerling, Gauss was elated over this process and mentioned that the computations could be undertaken even when a person is half asleep or thinking of more important things.

The linear systems that Gauss had to solve were strongly diagonally dominant and for that reason he could observe fast convergence. The Gauss–Seidel iteration process is much too slow for the very large linear systems that we see in many applications. For this reason there has been much research into faster methods and we will see the results of this later in this text.

1.2 Further arguments for iterative methods

For the solution of a linear system $Ax = b$, with A a nonsingular n by n matrix, we have the choice between direct and iterative methods.

The usual pro-arguments for iterative methods are based on economy of computer storage and (sometimes) CPU time. On the con side, it should be noted that the usage of iterative methods requires some expertise. If CPU-time and computer storage are not really at stake, then it would be unwise to consider iterative methods for the solution of a given linear system. The question remains whether there are situations where iterative solution methods are really preferable. In this section I will try to substantiate the pro-arguments; the con-arguments will appear in my more detailed presentation of iterative methods. I hope that the reader will feel sufficiently familiar, after reading these notes, with some of the more popular iterative methods in order to make a proper choice for the solving of classes of linear systems.

Dense linear systems, and sparse systems with a suitable nonzero structure, are most often solved by a so-called direct method, such as Gaussian elimination. A direct method leads, in the absence of rounding errors, to the exact solution of the given linear system in a finite and fixed amount of work. Rounding errors can be handled fairly well by pivoting strategies. Problems arise when the direct solution scheme becomes too expensive for the task. For instance, the elimination steps in Gaussian elimination may cause some zero entries of a sparse matrix to become nonzero entries, and nonzero entries require storage as well as CPU time. This is what may make Gaussian elimination, even with strategies for the reduction of the so-called fill-in, expensive.

In order to get a more quantitative impression of this, we consider a sparse system related to discretization of a second order PDE over a (not necessarily regular) grid, with about m unknowns per dimension. Think, for instance, of a finite element discretization over an irregular grid [159]. In a 3D situation this leads typically to a bandwidth $\sim n^{\frac{2}{3}}$ ($\approx m^2$ and $m^3 \approx n$, where $1/m$ is the (average) gridsize).

Gaussian elimination is carried out in two steps: first the matrix A is factored into a lower triangular matrix L, and an upper triangular matrix U (after suitable permutations of rows and columns):

$$A = LU.$$

When taking proper account of the band structure, the number of flops is then usually $\mathcal{O}(nm^4) \sim n^{2\frac{1}{3}}$ [98, 67]. We make the caveat 'usually', because it may happen that fill-in is very limited when the sparsity pattern of the matrix is special.

For 2D problems the bandwidth is $\sim n^{\frac{1}{2}}$, so that the number of flops for a direct method then varies with n^2.

Then, in the second step, we have to solve x from $LUx = b$, which, again, is done in two steps:

(a) first solve y from $Ly = b$,
(b) then solve x from $Ux = y$.

The LU factorization is the expensive part of the computational process; the solution of the two triangular systems is usually a minor cost item. If many systems with different right-hand sides have to be solved, then the matrix has to be factored only once, after which the cost for solving each system will vary with $n^{\frac{5}{3}}$ for 3D problems, and with $n^{\frac{3}{2}}$ for 2D problems.

In order to be able to quantify the amount of work for iterative methods, we have to be a little more specific. Let us assume that the given matrix is symmetric positive definite, in which case we may use the Conjugate Gradient (CG) method. The error reduction per iteration step of CG is $\sim \frac{\sqrt{\kappa}-1}{\sqrt{\kappa}+1}$, with $\kappa = \|A\|_2\|A^{-1}\|_2$ [44, 8, 98].

For discretized second order PDEs over grids with gridsize $\frac{1}{m}$, it can be shown that $\kappa \sim m^2$ (see, for instance, [159]). Hence, for 3D problems we have that $\kappa \sim n^{\frac{2}{3}}$, and for 2D problems: $\kappa \sim n$. In order to have an error reduction by a factor of ϵ, the number j of iteration steps must satisfy

$$\left(\frac{1-\frac{1}{\sqrt{\kappa}}}{1+\frac{1}{\sqrt{\kappa}}}\right)^j \approx \left(1-\frac{2}{\sqrt{\kappa}}\right)^j \approx \mathrm{e}^{-\frac{2j}{\sqrt{\kappa}}} < \epsilon.$$

For 3D problems, it follows that

$$j \sim -\frac{\log \epsilon}{2}\sqrt{\kappa} \approx -\frac{\log \epsilon}{2} n^{\frac{1}{3}},$$

whereas for 2D problems,

$$j \approx -\frac{\log \epsilon}{2} n^{\frac{1}{2}}.$$

If we assume the number of flops per iteration to be $\sim fn$ (f stands for the average number of nonzeros per row of the matrix and the overhead per unknown introduced by the iterative scheme), then the required number of flops for a reduction of the initial error with ϵ is

(a) for 3D problems: $\sim -fn^{\frac{4}{3}} \log \epsilon$, and
(b) for 2D problems: $\sim -fn^{\frac{3}{2}} \log \epsilon$.

f is typically a modest number, say of order 10–15.

From comparing the flops counts for the direct scheme with those for the iterative CG method we conclude that the CG method may be preferable if we have to solve one system at a time, and if n is large, or f is small, or ϵ is modest.

If we have to solve many systems $Ax = b_k$ with different right-hand sides b_k, and if we assume their number to be so large that the costs for constructing the LU factorization of A is relatively small per system, then it seems likely that direct methods will be more efficient for 2D problems. For 3D problems this is unlikely, because the flops counts for the two triangular solves associated with a direct solution method are proportional to $n^{\frac{5}{3}}$, whereas the number of flops for the iterative solver (for the model situation) varies in the same way as $n^{\frac{4}{3}}$.

1.3 An example

The above arguments are quite nicely illustrated by observations made by Horst Simon [173]. He predicted that by now we will have to solve routinely linear problems with some 5×10^9 unknowns. From extrapolation of the CPU times observed for a characteristic model problem, he estimated the CPU time for the most efficient direct method as 520 040 years, provided that the computation can be carried out at a speed of 1 TFLOPS.

On the other hand, the extrapolated guess for the CPU time with preconditioned conjugate gradients, still assuming a processing speed of 1 TFLOPS, is 575 seconds. As we will see, the processing speed for iterative methods may be a factor lower than for direct methods, but, nevertheless, it is obvious that the differences in CPU time requirements are gigantic. The ratio of the two times is of order n, just as we might have expected from our previous arguments.

Also the requirements for memory space for the iterative methods are typically smaller by orders of magnitude. This is often the argument for the use of iterative methods in 2D situations, when flops counts for both classes of methods are more or less comparable.

Remarks:

- With suitable preconditioning we may have $\sqrt{\kappa} \sim n^{\frac{1}{6}}$ and the flops count is then roughly proportional to

$$-fn^{\frac{7}{6}} \log \epsilon,$$

  see, e.g., [105, 10].
- Special methods may even be (much) faster for special problems: Fast Poisson Solvers [35, 184], multigrid [111, 220]. For more general problems, we see combinations of these methods with iterative schemes. For instance, iterative schemes can be used as smoothers for multigrid, or a multigrid cycle for an approximating regular problem may be used as a preconditioner for an iterative method for an irregular problem. Also, preconditioners have been designed with multigrid-like properties.
- For matrices that are not positive definite symmetric the situation can be more problematic: it is often difficult to find the proper iterative method or a suitable preconditioner. However, for methods related in some sense to CG, like GMRES, QMR, TFQMR, Bi-CG, CGS, and Bi-CGSTAB, we often see that the flops counts are similar to those for CG.
- Iterative methods can be attractive even when the matrix is dense. Again, in the positive definite symmetric case, if the condition number is $n^{2-2\varepsilon}$ then, since the amount of work per iteration step is $\sim n^2$ and the number of iteration steps $\sim n^{1-\varepsilon}$, the total work estimate is roughly proportional to $n^{3-\varepsilon}$, and this is asymptotically less than the amount of work for Cholesky's method (the symmetric positive definite variant of Gaussian elimination), which varies with $\sim n^3$. This says that the condition number has to be less than n^2 in order to make iterative methods potentially competitive for dense matrices.
- In many situations the condition number tells only part of the story. Methods like CG can be a good deal faster than the condition number predicts. This happens, for instance, when the eigenvalues of A have relatively big gaps at the lower end of the spectrum (see Section 5.3).

1.4 Performance aspects

Now the question remains – how well can iterative methods take advantage of modern computer architectures? From Dongarra's LINPACK benchmark [59]

Table 1.2. Speed in megaflops for 50 Iterations of ICCG and CG

Machine	Peak performance	Optimized ICCG	Scaled CG
NEC SX-3/22 (2.9 ns)	2750	607	1124
CRAY Y-MP C90 (4.2 ns)	952	444	737
CRAY 2 (4.1 ns)	500	96	149
IBM 9000 Model 820	444	40	75
IBM 9121 (15 ns)	133	11	25
DEC Vax/9000 (16 ns)	125	10	17
IBM RS/6000-550 (24 ns)	81	18	21
CONVEX C3210	50	16	19
Alliant FX2800	40	2	3

Table 1.3. Performances in megaflops for processors

Processor	Peak performance	CG	GMRES	ILU
EV6	759	285	216	163
Athlon 600 MHz	154	43	44	34
SGI Origin	106	70	71	57
ALPHA 533 MHz	81	45	40	33
IBM 375 MHz	606	254	209	120
SUN 296 MHz	154	57	37	34
R1200 270 MHz	155	52	78	62
PPC G4 450 MHz	198	45	38	31
Pentium III 550 MHz	96	37	39	27

it may be concluded that the solution of a dense linear system can (in principle) be computed with computational speeds close to peak speeds on most computers. This is already the case for systems of, say, order 50 000 on parallel machines with as many as 1024 processors.

In sharp contrast to the dense case are computational speeds reported in [63] for the preconditioned as well as the unpreconditioned conjugate gradient methods (ICCG and CG, respectively). I list some of these results for classical vector computers, for regular sparse systems from 7-point stencil discretizations of 3D elliptic PDEs, of order $n = 10^6$, in Table 1.2. We see that, especially for ICCG, the performance stays an order of magnitude below the theoretical peak of most machines.

A benchmark for representative components of iterative methods is proposed in [62]. In this benchmark the asymptotic speeds, that is the megaflops rates for large values of the order of the linear system, are computed for the standard unpreconditioned Conjugate Gradients method and the unpreconditioned GMRES(20) method. The performance of preconditioners, necessary in order to speed up the convergence, is measured separately. When a preconditioner is used, the overall performance of the iterative process is typically closer to the performance of the preconditioning part. In Table 1.3 we list some asymptotic performances for Conjugate Gradients (CG), GMRES(20) (GMRES), and the popular preconditioner ILU, for more modern processors. The situation for these processors is certainly somewhat better than for the vector processors. However, note that high performances are now sought by coupling large numbers of these processors (massively parallel computing). Parallelism in preconditioned iterative methods is not a trivial matter. I will come back to this in my discussion on preconditioning techniques.

2

Mathematical preliminaries

In this chapter I have collected some basic notions and notations that will be used throughout this book.

2.1 Matrices and vectors

We will look at linear systems $Ax = b$, where A is usually an n by n matrix:

$$A \in \mathbb{R}^{n \times n}.$$

The elements of A will be denoted as $a_{i,j}$. The vectors $x = (x_1, x_2, \ldots, x_n)^T$ and b belong to the linear space $\mathbb{R}^n$. Sometimes we will admit complex matrices $A \in \mathbb{C}^{n \times n}$ and vectors $x, b \in \mathbb{C}^n$, but that will be explicitly mentioned.

Over the space $\mathbb{R}^n$ we will use the Euclidean inner product between two vectors x and y:

$$x^T y = \sum_{i=1}^{n} x_i y_i,$$

and for $v, w \in \mathbb{C}^n$ we use the standard complex inner product:

$$v^H w = \sum_{i=1}^{n} \bar{v}_i w_i.$$

These inner products lead to the 2-norm or Euclidean length of a vector

$$\|x\|_2 = \sqrt{x^T x} \text{ for } x \in \mathbb{R}^n,$$

$$\|v\|_2 = \sqrt{v^H v} \text{ for } v \in \mathbb{C}^n.$$

With these norms we can associate a 2 norm for matrices: for $A \in \mathbb{R}^{n\times n}$, its associated 2-norm $\|A\|_2$ is defined as

$$\|A\|_2 = \sup_{y\in\mathbb{R}^n, y\neq 0} \frac{\|Ay\|_2}{\|y\|_2},$$

and similarly in the complex case, using the complex inner product.

The associated matrix norms are convenient, because they can be used to bound products. For $A \in \mathbb{R}^{n\times k}$, $B \in \mathbb{R}^{k\times m}$, we have that

$$\|AB\|_2 \leq \|A\|_2 \|B\|_2,$$

in particular

$$\|Ax\|_2 \leq \|A\|_2 \|x\|_2.$$

The inverse of a nonsingular matrix A is denoted as A^{-1}. Particularly useful is the condition number of a square nonsingular matrix A, defined as

$$\kappa_2(A) = \|A\|_2 \|A^{-1}\|_2.$$

The condition number is used to characterize the sensitivity of the solution x of $Ax = b$ with respect to perturbations in b and A. For perturbed systems we have the following theorem.

Theorem 2.1. *[98, Theorem 2.7.2] Suppose*

$$Ax = b \quad A \in \mathbb{R}^{n\times n}, 0 \neq b \in \mathbb{R}^n$$

$$(A + \Delta A)y = b + \Delta b \quad \Delta A \in \mathbb{R}^{n\times n}, \Delta b \in \mathbb{R}^n,$$

with $\|\Delta A\|_2 \leq \epsilon \|A\|_2$ *and* $\|\Delta b\|_2 \leq \epsilon \|b\|_2$.

If $\epsilon\kappa_2(A) = r < 1$, *then* $A + \Delta A$ *is nonsingular and*

$$\frac{\|y - x\|_2}{\|x\|_2} \leq \frac{2\epsilon}{1 - r}\kappa_2(A).$$

With the superscript ^{T} we denote the transpose of a matrix (or vector): for $A \in \mathbb{R}^{n\times k}$, the matrix $B = A^T \in \mathbb{R}^{k\times n}$ is defined by

$$b_{i,j} = a_{j,i}.$$

If $E \in \mathbb{C}^{n \times k}$ then the superscript ^{H} is used to denote its complex conjugate $F = E^H$, defined as

$$f_{i,j} = \bar{e}_{j,i}.$$

Sometimes the superscript ^{T} is used for complex matrices in order to denote the transpose of a complex matrix.

The matrix A is symmetric if $A = A^T$, and $B \in \mathbb{C}^{n \times n}$ is Hermitian if $B = B^H$. Hermitian matrices have the attractive property that their spectrum is real. In particular, Hermitian (or symmetric real) matrices that are positive definite are attractive, because they can be solved rather easily by proper iterative methods (the CG method).

A Hermitian matrix $A \in \mathbb{C}^{n \times n}$ is positive definite if $x^H Ax > 0$ for all $0 \neq x \in \mathbb{C}^n$. A positive definite Hermitian matrix has only positive real eigenvalues.

In the context of preconditioning, the notion of an M-matrix is useful. Some popular preconditioners can be proven to exist when A is an M-matrix.

Definition 2.1. *A nonsingular $A \in \mathbb{R}^{n \times n}$ is an M-matrix if $a_{i,j} \leq 0$ for $i \neq j$ and $A^{-1} \geq 0$.*

With $A \geq 0$ we denote the situation that the inequality holds for all elements of A. The M-matrix property can be proven to hold for important classes of discretized PDEs. It is an important property for iteration methods that are based on (regular) splittings of the matrix A (for details on this see [212]).

We will encounter some special matrix forms, in particular tridiagonal matrices and (upper) Hessenberg matrices. The matrix $T = (t_{i,j}) \in \mathbb{R}^{n \times m}$ will be called tridiagonal if all elements for which $|i - j| > 1$ are zero. It is called upper Hessenberg if all elements for which $i > j + 1$ are zero. In the context of Krylov subspaces, these matrices are often $k + 1$ by k and they will then be denoted as $T_{k+1,k}$.

2.2 Eigenvalues and eigenvectors

For purposes of analysis it is often helpful or instructive to transform a given matrix to an easier form, for instance, diagonal or upper triangular form.

The easiest situation is the symmetric case: for a real symmetric matrix, there exists an orthogonal matrix $Q \in \mathbb{R}^{n \times n}$, so that $Q^T AQ = D$, where $D \in \mathbb{R}^{n \times n}$ is a diagonal matrix. The diagonal elements of D are the eigenvalues of A, the columns of Q are the corresponding eigenvectors of A. Note that the eigenvalues and eigenvectors of A are all real.

If $A \in \mathbb{C}^{n \times n}$ is Hermitian ($A = A^H$) then there exist $Q \in \mathbb{C}^{n \times n}$ and a diagonal matrix $D \in \mathbb{R}^{n \times n}$ so that $Q^H Q = I$ and $Q^H A Q = D$. This means that the eigenvalues of a Hermitian matrix are all real, but its eigenvalues may be complex.

Unsymmetric matrices do not in general have an orthonormal set of eigenvectors, and may not have a complete set of eigenvectors, but they can be transformed unitarily to Schur form:

$$Q^* A Q = R,$$

in which R is upper triangular. In fact, the symmetric case is a special case of this Schur decomposition, since a symmetric triangular matrix is clearly diagonal. Apart from the ordering of the eigenvalues along the diagonal of R and the sign of each column of Q, the matrix Q is unique.

If the matrix A is complex, then the matrices Q and R may also be complex. However, they may be complex even when A is real unsymmetric. It may then be advantageous to work in real arithmetic. This can be realized because of the existence of the *real Schur decomposition*. If $A \in \mathbb{R}^{n \times n}$ then it can be transformed with an orthonormal $Q \in \mathbb{R}^{n \times n}$ as

$$Q^T A Q = \widetilde{R},$$

with

$$\widetilde{R} = \begin{bmatrix} \widetilde{R}_{1,1} & \widetilde{R}_{1,2} & \cdots & \widetilde{R}_{1,k} \\ 0 & \widetilde{R}_{2,2} & \cdots & \widetilde{R}_{2,k} \\ \vdots & \vdots & \ddots & \vdots \\ 0 & 0 & \cdots & \widetilde{R}_{k,k} \end{bmatrix} \in \mathbb{R}^{n \times n}.$$

Each $\widetilde{R}_{i,i}$ is either 1 by 1 or a 2 by 2 (real) matrix having complex conjugate eigenvalues. For a proof of this see [98, Chapter 7.4.1]. This form of $\widetilde{R}$ is referred to as an upper *quasi-triangular matrix*.

If all eigenvalues are distinct then there exists a nonsingular matrix X (in general not orthogonal) that transforms A to diagonal form:

$$X^{-1} A X = D.$$

A general matrix can be transformed to Jordan form with a nonsingular X:

$$X^{-1} A X = \operatorname{diag}(J_1, J_2, \ldots, J_k),$$

where

$$J_i = \begin{bmatrix} \lambda_i & 1 & 0 & \cdots & 0 \\ 0 & \lambda_i & \ddots & & \vdots \\ & \ddots & \ddots & \ddots & \\ \vdots & & \ddots & \ddots & 1 \\ 0 & \cdots & & 0 & \lambda_i \end{bmatrix}.$$

If there is a J_i with dimension greater than 1 then the matrix A is defective. In this case A does not have a complete set of independent eigenvectors. In numerical computations we may argue that small perturbations lead to different eigenvalues and hence that it will be unlikely that A has a true Jordan form in actual computation. However, if A is close to a matrix with a nontrivial Jordan block, then this is reflected by a (severely) ill-conditioned eigenvector matrix X.

Matrices $A \in \mathbb{C}^{n\times n}$ that satisfy the property $A^H A = AA^H$ are called *normal*. Normal matrices also have a complete orthonormal eigensystem. For such matrices the distribution of the eigenvalues can help to explain (local) phenomena in the convergence behaviour of some methods. For unsymmetric matrices that are not normal, the eigenvalues are often insufficient for a detailed analysis. In some situations the convergence behaviour can be analysed partly with the so-called *field of values*.

Definition 2.2. *The* field of values $\mathcal{F}(A)$ *is defined as*

$$\mathcal{F}(A) = \{z^H Az \mid z^H z = 1\}.$$

We will also encounter eigenvalues that are called *Ritz values*. For simplicity, we will introduce them here for the real case. The subspace methods that will be discussed in this book are based on the identification of good solutions from certain low-dimensional subspaces $\mathcal{V}^k \subset \mathbb{R}^n$, where $k \ll n$ denotes the dimension of the subspace. If $V_k \in \mathbb{R}^{n\times k}$ denotes an orthogonal basis of $\mathcal{V}^k$ then the operator $H_k = V_k^T A V_k \in \mathbb{R}^{k\times k}$ represents the projection of A onto V_k. Assume that the eigenvalues and eigenvectors of H_k are represented as

$$H_k s_j^{(k)} = \theta_j^{(k)} s_j^{(k)},$$

the $\theta_j^{(k)}$ is called a *Ritz value* of A with respect to $\mathcal{V}^k$ and $V_k s_j^{(k)}$ is its corresponding *Ritz vector*. For a thorough discussion of Ritz values and Ritz vectors see, for instance, [155, 165, 182, 203].

For some methods we will see that *Harmonic Ritz values* play a role. Let W_k denote an orthogonal basis for the subspace $A\mathcal{V}^k$ then the Harmonic Ritz values of A with respect to that subspace are the inverses of the eigenvalues of the projection Z_k of A^{-1}:

$$Z_k = W_k^T A^{-1} W_k.$$

The importance of the (Harmonic) Ritz values is that they can be viewed as approximations for eigenvalues of A. Often they represent, even for modest values of k, very accurate approximations for some eigenvalues. The monitoring of Ritz values, which can often be easily obtained as an inexpensive side product of the iteration process, reveals important information on the iteration process and on the (preconditioned) matrix A.

3
Basic iteration methods

3.1 Introduction

The idea behind iterative methods is to replace the given system by some nearby system that can be more easily solved. That is, instead of $Ax = b$ we solve the simpler system $Kx_0 = b$ and take x_0 as an approximation for x. The iteration comes from the systematic way in which the approximation can be improved. Obviously, we want the correction z that satisfies

$$A(x_0 + z) = b.$$

This leads to a new linear system

$$Az = b - Ax_0.$$

Again, we solve this system by a nearby system, and most often K is again taken:

$$Kz_0 = b - Ax_0.$$

This leads to the new approximation $x_1 = x_0 + z_0$. The correction procedure can now be repeated for x_1, and so on, which gives us an iterative method. In some iteration methods we select a cycle of different approximations K, as, for instance, in ADI. In such cases the approximation for x after one cycle can be regarded as being obtained from the approximation prior to the cycle with an implicitly constructed K that represents the full cycle. This observation is of importance for the construction of preconditioners.

For the basic iteration, introduced above, it follows that

$$\begin{aligned} x_{i+1} &= x_i + z_i \\ &= x_i + K^{-1}(b - Ax_i) \\ &= x_i + \widetilde{b} - \widetilde{A}x_i, \end{aligned} \tag{3.1}$$

with $\widetilde{b} = K^{-1}b$ and $\widetilde{A} = K^{-1}A$. We use K^{-1} only for notational purposes; we (almost) never compute inverses of matrices explicitly. When we speak of $K^{-1}b$, we mean the vector $\widetilde{b}$ that is solved from $K\widetilde{b} = b$, and in the same way for $K^{-1}Ax_i$.

The formulation in (3.1) can be interpreted as the basic iteration for the preconditioned linear system

$$\widetilde{A}x = \widetilde{b}, \tag{3.2}$$

with approximation $K = I$ for $\widetilde{A} = K^{-1}A$.

In order to simplify our formulas, we will from now on assume that if we have some preconditioner K, we apply our iterative schemes to the (preconditioned) system (3.2), and we will skip the superscript $\widetilde{\ }$. This means that we iterate for $Ax = b$ with approximation $K = I$ for A. In some cases it will turn out to be more convenient to incorporate the preconditioner explicitly in the iteration scheme, but that will be clear from the context.

We have thus arrived at the well-known Richardson iteration:

$$x_{i+1} = b + (I - A)x_i = x_i + r_i, \tag{3.3}$$

with the residual $r_i = b - Ax_i$.

Because relation (3.3) contains x_i as well as r_i, it cannot easily be analysed. Multiplication by $-A$ and adding b gives

$$b - Ax_{i+1} = b - Ax_i - Ar_i$$

or

$$r_{i+1} = (I - A)r_i \tag{3.4}$$

$$\begin{aligned} &= (I - A)^{i+1}r_0 \\ &= P_{i+1}(A)r_0. \end{aligned} \tag{3.5}$$

In terms of the error $x - x_i$, we get

$$A(x - x_{i+1}) = P_{i+1}(A)A(x - x_0),$$

so that, for nonsingular A:

$$x - x_{i+1} = P_{i+1}(A)(x - x_0).$$

In these expressions P_{i+1} is a (special) polynomial of degree $i + 1$. Note that $P_{i+1}(0) = 1$.
The expressions (3.4) and (3.5) lead to interesting observations. From (3.4) we conclude that

$$\|r_{i+1}\| \leq \|I - A\|\|r_i\|,$$

which shows that we have guaranteed convergence for all initial r_0 if $\|I - A\| < 1$. This puts restrictions on the preconditioner (remember that A represents the preconditioned matrix). We will later see that for the convergence of more advanced iterative schemes, we may drop the restriction $\|I - A\| < 1$.

Equation (3.5) is also of interest, because it shows that all residuals can be expressed in terms of powers of A times the initial residual. This observation will be crucial for the derivation of methods like the Conjugate Gradients method. It shows something more. Let us assume that A has n eigenvectors w_j, with corresponding eigenvectors λ_j. Then we can express r_0 in terms of the eigenvector basis as

$$r_0 = \sum_{j=1}^{n} \gamma_j w_j,$$

and we see that

$$r_i = P_i(A)r_0 = \sum_{j=1}^{n} \gamma_j P_i(\lambda_j) w_j.$$

This formula shows that the error reduction depends on how well the polynomial P_i damps the initial error components. It would be nice if we could construct iterative methods for which the corresponding error reduction polynomial P_i has better damping properties than for the standard iteration (3.3).

From now on we will also assume that $x_0 = 0$ to simplify future formulas. This does not mean a loss of generality, because the situation $x_0 \neq 0$ can be transformed with a simple shift to the system

$$Ay = b - Ax_0 = \bar{b} \tag{3.6}$$

for which obviously $y_0 = 0$.

With the simple Richardson iteration, we can proceed in different ways. One way is to include iteration parameters, for instance, by computing x_{i+1} as

$$x_{i+1} = x_i + \alpha_i r_i. \tag{3.7}$$

This leads to the error reduction formula

$$r_{i+1} = (I - \alpha_i A) r_i.$$

It follows that the error reduction polynomial P_i can be expressed in this case as

$$P_i = \prod_{j=1}^{i} (I - \alpha_j A).$$

Apparently, we can now construct methods for which the corresponding error polynomials have better damping properties. For instance, if the eigenvalues λ_j are all in the real interval $[a, b]$, with $a > 0$, then we can select the α_j as the zeros of a Chebyshev polynomial, shifted from $[-1, 1]$ to $[a, b]$ and scaled so that $P_i(0) = 1$, which leads to the *Chebyshev iteration method.* An obvious criticism could be that we have to select different sets of iteration parameters for different values of i (because all Chebyshev polynomials have different zeros), but it turns out that the recurrence relations for Chebyshev polynomials can be exploited to derive recurrence relations for the corresponding x_i. For further details on this see [212, Chapter 5] or [98, Chapter 10.1.5].

An important consequence of this polynomial damping interpretation is that it is no longer necessary that $I - A$ has all its eigenvalues inside the unit ball. The eigenvalues may, in principle, be anywhere as long as we have a chance to construct, implicitly, iteration polynomials that damp the unwanted error components.

A seemingly different approach is to save all approximations x_i and to try to recombine them into something better. This may seem awkward, but after some reflection we see that all approximations can be expressed in polynomial form, and hence also any possible combination. This shows that the Chebyshev iteration is a member of this class, but we may hope that the approach leads to optimal methods; optimal in a sense to be defined later.

First we have to identify the subspace in which the successive approximate solutions are located. By repeating the simple Richardson iteration, we observe that

$$x_{i+1} = r_0 + r_1 + r_2 + \cdots + r_i \tag{3.8}$$

$$= \sum_{j=0}^{i} (I - A)^j r_0 \tag{3.9}$$

$$\in \text{span}\{r_0, Ar_0, \ldots, A^i r_0\} \tag{3.10}$$

$$\equiv K^{i+1}(A; r_0). \tag{3.11}$$

The m-dimensional space spanned by a given vector v, and increasing powers of A applied to v, up to the $(m-1)$-th power, is called the m dimensional Krylov subspace, generated with A and v, denoted by $K^m(A; v)$.

Apparently, the Richardson iteration, as it proceeds, delivers elements of Krylov subspaces of increasing dimension. This is also the case for the Richardson iteration (3.7) with parameters. Including local iteration parameters in the iteration would lead to other elements of the same Krylov subspaces. Let us still write such an element as x_{i+1}. Since $x_{i+1} \in K^{i+1}(A; r_0)$, we have that

$$x_{i+1} = Q_i(A) r_0,$$

with Q_i an arbitrary polynomial of degree i. It follows that

$$r_{i+1} = b - Ax_{i+1} = (I - AQ_i(A))r_0 = \widetilde{P}_{i+1}(A) r_0, \tag{3.12}$$

with, just as in the standard Richardson iteration, $\widetilde{P}_{i+1}(0) = 1$. The standard Richardson iteration is characterized by the polynomial $P_{i+1}(A) = (I - A)^{i+1}$.

The consequence is that if we want to make better combinations of the generated approximations, then we have to explore the Krylov subspace.

3.2 The Krylov subspace approach

Methods that attempt to generate better approximations from the Krylov subspace are often referred to as Krylov subspace methods. Because optimality usually refers to some sort of projection, they are also called Krylov projection methods. The Krylov subspace methods, for identifying suitable $x \in \mathcal{K}^k(A; r_0)$, can be distinguished in four different classes (we will still assume that $x_0 = 0$):

(1) The *Ritz–Galerkin approach*: Construct the x_k for which the residual is orthogonal to the current subspace: $b - Ax_k \perp \mathcal{K}^k(A; r_0)$.
(2) The *minimum norm residual approach*: Identify the x_k for which the Euclidean norm $\|b - Ax_k\|_2$ is minimal over $\mathcal{K}^k(A; r_0)$.

(3) The *Petrov–Galerkin approach*: Find an x_k so that the residual $b - Ax_k$ is orthogonal to some other suitable k-dimensional subspace.
(4) The *minimum norm error approach*: Determine x_k in $A^T \mathcal{K}^k(A^T; r_0)$ for which the Euclidean norm $\|x_k - x\|_2$ is minimal.

The Ritz–Galerkin approach leads to well-known methods such as Conjugate Gradients, the Lanczos method, FOM, and GENCG. The minimum norm residual approach leads to methods such as GMRES, MINRES, and ORTHODIR. The main disadvantage of these two approaches is that, for most unsymmetric systems, they lead to long and therefore expensive recurrence relations for the approximate solutions. This can be relieved by selecting other subspaces for the orthogonality condition (the Galerkin condition). If we select the k-dimensional subspace in the third approach as $\mathcal{K}^k(A^T; s_0)$, we then obtain the Bi-CG and QMR methods, and these methods indeed work with short recurrences. The fourth approach is not so obvious, but for $A = A^T$ it already becomes more natural. In this case it leads to the SYMMLQ method [153]. For the unsymmetric case it leads to the less well-known GMERR methods [218, 219]. Hybrids of these approaches have been proposed, for instance CGS, Bi-CGSTAB, Bi-CGSTAB(ℓ), TFQMR, FGMRES, and GMRESR.

The choice of a method is a delicate problem. If the matrix A is symmetric positive definite, then the choice is easy: Conjugate Gradients. For other types of matrices the situation is very diffuse. GMRES, proposed in 1986 by Saad and Schultz [169], is the most robust method, but in terms of work per iteration step it is also relatively expensive. Bi-CG, which was suggested by Fletcher in 1976 [83], is a relatively inexpensive alternative, but it has problems with respect to convergence: the so-called breakdown situations. This aspect has received much attention in the literature. Parlett et al. [156] introduced the notion of look-ahead in order to overcome breakdowns and this was further perfected by Freund, Gutknecht and Nachtigal [89], and by Brezinski and Redivo Zaglia [32]. The theory for this look-ahead technique was linked to the theory of Padé approximations by Gutknecht [108]. Other contributions to overcome specific breakdown situations were made by Bank and Chan [17], and Fischer [81]. I discuss these approaches in the sections on Bi-CG and QMR.

The development of hybrid methods started with CGS, published in 1989 by Sonneveld [180], and was followed by Bi-CGSTAB, by van der Vorst in 1992 [201], and others. The hybrid variants of GMRES: Flexible GMRES and GMRESR, in which GMRES is combined with some other iteration scheme, were proposed in the mid-1990s.

A nice overview of Krylov subspace methods, with focus on Lanczos-based methods, is given in [88]. Simple algorithms and unsophisticated software for some of these methods are provided in the 'Templates' book [20]. This book is complemented, with respect to theory, by the very elegant textbook [101], authored by Greenbaum. Iterative methods with much attention to various forms of preconditioning have been described in [10]. Another book on iterative methods was published by Saad [168]; it is very algorithm oriented, with, of course, a focus on GMRES and preconditioning techniques, for instance threshold ILU, ILU with pivoting, and incomplete LQ factorizations. A nice introduction to Krylov subspace methods, viewed from the standpoint of polynomial methods, can be found in [82]. An annotated entrance to the vast literature on preconditioned iterative methods is given in [33].

3.3 The Krylov subspace

In order to identify the approximations corresponding to the four different approaches, we need a suitable basis for the Krylov subspace; one that can be extended in a meaningful way for subspaces of increasing dimension. The obvious basis r_0, $Ar_0, \ldots, A^{i-1}r_0$ for $\mathcal{K}^i(A; r_0)$, is not very attractive from a numerical point of view, since the vectors $A^j r_0$ point more and more in the direction of the dominant eigenvector for increasing j (the power method!), and hence the basis vectors become dependent in finite precision arithmetic. It does not help to compute this nonorthogonal generic basis first and to orthogonalize it afterwards. The result would be that we have orthogonalized a very ill-conditioned set of basis vectors, which is numerically still not an attractive situation.

We now derive an orthogonal basis that, in exact arithmetic, spans the Krylov subspace. For this we follow ideas from [182, Chapter 4.3]. We start with the generic basis for $\mathcal{K}^{i+1}(A; r_0)$ and we denote the basis vectors by u_j:

$$u_j = A^{j-1} r_0.$$

We define the n by j matrix U_j as the matrix with columns $u_1, \ldots, u_j$. The connection between A and U_i is left as an exercise.

Exercise 3.1. *Show that*

$$AU_i = U_i B_i + u_{i+1} e_i^T, \tag{3.13}$$

with e_i the i-th canonical basis vector in $\mathbb{R}^i$, and B_i an i by i matrix with $b_{j+1,j} = 1$ and all other elements zero.

The next step is to decompose U_i, still in exact arithmetic, as

$$U_i = Q_i R_i,$$

with $Q_i^T Q_i = I$ and R_i upper triangular. Then, with (3.13), it follows that

$$A Q_i R_i = Q_i R_i B_i + u_{i+1} e_i^T,$$

or

$$\begin{aligned} A Q_i &= Q_i R_i B_i R_i^{-1} + u_{i+1} e_i^T R_i^{-1} \\ &= Q_i \widetilde{H}_i + u_{i+1} e_i^T R_i^{-1} && (3.14) \\ &= Q_i \widetilde{H}_i + \frac{1}{r_{i,i}} u_{i+1} e_i^T. && (3.15) \end{aligned}$$

Exercise 3.2. *Show that $\widetilde{H}_i$ is an upper Hessenberg matrix.*

We can also decompose U_{i+1} as $U_{i+1} = Q_{i+1} R_{i+1}$, and if we write the last column of R_{i+1} as $(\widetilde{r}, r_{i+1,i+1})^T$, that is

$$R_{i+1} = \begin{pmatrix} R_i & \widetilde{r} \\ 0 & r_{i+1,i+1} \end{pmatrix},$$

then it follows that

$$u_{i+1} = Q_i \widetilde{r} + r_{i+1,i+1} q_{i+1}.$$

In combination with (3.14), this gives

$$\begin{aligned} A Q_i &= Q_i (\widetilde{H}_i + \frac{1}{r_{i,i}} \widetilde{r} e_i^T) + \frac{r_{i+1,i+1}}{r_{i,i}} q_{i+1} e_i^T \\ &= Q_i H_i + \alpha q_{i+1} e_i^T. && (3.16) \end{aligned}$$

From this expression we learn at least two things: first

$$Q_i^T A Q_i = H_i, \tag{3.17}$$

with H_i upper Hessenberg, and second

$$q_{i+1}^T A q_i = \alpha,$$

which, with $Q_{i+1}^T A Q_{i+1} = H_{i+1}$, leads to $\alpha = h_{i+1,i}$.

The implicit Q theorem [98, Theorem 7.4.2] states that the orthogonal Q that reduces A to upper Hessenberg form is uniquely determined by $q_1 = \frac{1}{\|r_0\|} r_0$, except for signs (that is, q_j may be multiplied by -1). The orthogonality of the q_j basis gives us excellent opportunities to compute this basis in finite precision arithmetic.

Arnoldi [6] proposed to compute the orthogonal basis as follows. In fact, with Arnoldi's procedure we compute in a straightforward manner the columns of Q_i and the elements of H_i. Start with $v_1 \equiv r_0/\|r_0\|_2$. Then compute Av_1, make it orthogonal to v_1 and normalize the result, which gives v_2. The general procedure is as follows. Assuming we already have an orthonormal basis $v_1, \ldots, v_j$ for $K^j(A; r_0)$, this basis is expanded by computing $t = Av_j$ and by orthonormalizing this vector t with respect to $v_1, \ldots, v_j$. In principle the orthonormalization process can be carried out in different ways, but the most commonly used approach is then the modified Gram–Schmidt procedure [98].
This leads to an algorithm for the creation of an orthonormal basis for $\mathcal{K}^m(A; r_0)$, as in Figure 3.1. It is easily verified that $v_1, \ldots, v_m$ form an orthonormal basis for $\mathcal{K}^m(A; r_0)$ (that is, if the construction does not terminate at a vector $t = 0$). The orthogonalization leads, in exact arithmetic, to the relation that we have seen before (cf. (3.16), but now expressed in terms of the v_j. Let V_j denote the matrix with columns v_1 up to v_j then it follows that

$$AV_{m-1} = V_m H_{m,m-1}. \tag{3.18}$$

The m by $m-1$ matrix $H_{m,m-1}$ is upper Hessenberg, and its elements $h_{i,j}$ are defined by the Arnoldi algorithm.

```
v_1 = r_0/||r_0||_2;
for j = 1, ..., m - 1
    t = A v_j;
    for i = 1, ..., j
        h_{i,j} = v_i^T t;
        t = t - h_{i,j} v_i;
    end;
    h_{j+1,j} = ||t||_2;
    v_{j+1} = t/h_{j+1,j};
end
```

Figure 3.1. Arnoldi's method with modified Gram–Schmidt orthogonalization.

From a computational point of view, this construction is composed of three basic elements: a matrix vector product with A, inner products, and vector updates. We see that this orthogonalization becomes increasingly expensive for increasing dimension of the subspace, since the computation of each $h_{i,j}$ requires an inner product and a vector update.

Note that if A is symmetric, then so is $H_{m-1,m-1} = V_{m-1}^T A V_{m-1}$, so that in this situation $H_{m-1,m-1}$ is tridiagonal. This means that in the orthogonalization process, each new vector has to be orthogonalized with respect to the previous two vectors only, since all other inner products vanish. The resulting three-term recurrence relation for the basis vectors of $K_m(A; r_0)$ is known as the *Lanczos method* [129] and some very elegant methods are derived from it. In this symmetric case the orthogonalization process involves constant arithmetical costs per iteration step: one matrix vector product, two inner products, and two vector updates.

3.3.1 A more accurate basis for the Krylov subspace

A more accurate implementation for the construction of an orthonormal basis, useful for ill-conditioned matrices A, was suggested by Walker [215]. He

```
v is a convenient starting vector
Select a value for κ, e.g., κ = .25
v_1 = v/||v||_2
for j = 1, ..., m − 1
    t = Av_j
    τ_in = ||t||_2
    for i = 1, ..., j
        h_{i,j} = v_i^* t
        t = t − h_{i,j} v_i
    end
    if ||t||_2/τ_in ≤ κ
        for i = 1, ..., j
            ρ = v_i^* t
            t = t − ρ v_i
            h_{i,j} = h_{i,j} + ρ
        end
    endif
    h_{j+1,j} = ||t||_2
    v_{j+1} = t/h_{j+1,j}
end
```

Figure 3.2. The Arnoldi Method with refined modified Gram–Schmidt.

suggested employing Householder reflections instead of the modified Gram–Schmidt orthogonalization procedure.

An alternative is to do two iterations with (modified) Gram–Schmidt if necessary. This works as follows. If we want to have a set of orthogonal vectors to almost working precision then we have to check, after the orthogonalization of a new vector with respect to the existing set, whether the resulting unnormalized vector is significantly smaller in norm than the new vector at the start of the orthogonalization step, say more than $\kappa < 1$ smaller. In that case we may have had cancellation effects, and once again we apply modified Gram–Schmidt. This is the basis for the refinement technique suggested in [50]. It leads to a set of vectors for which the mutual loss of orthogonality is limited to $1/\kappa$, in a relative sense. In the template in Figure 3.2, we incorporate this technique into the Arnoldi algorithm.

Note that, in exact arithmetic, the constants ρ in Figure 3.2 are equal to zero. It is easily verified that, in exact arithmetic, the $v_1, \ldots, v_m$ form an orthonormal basis for $\mathcal{K}^m(A; v)$ (that is, if the construction does not terminate at a vector $t = 0$).

4

Construction of approximate solutions

4.1 The Ritz–Galerkin approach

The Ritz–Galerkin conditions imply that $r_k \perp \mathcal{K}^k(A; r_0)$, and this is equivalent to

$$V_k^T(b - Ax_k) = 0.$$

Since $b = r_0 = \|r_0\|_2 v_1$, it follows that $V_k^T b = \|r_0\|_2 e_1$ with e_1 the first canonical unit vector in $\mathbb{R}^k$. With $x_k = V_k y$ we obtain

$$V_k^T A V_k y = \|r_0\|_2 e_1.$$

This system can be interpreted as the system $Ax = b$ projected onto the subspace $\mathcal{K}^k(A; r_0)$.

Obviously we have to construct the $k \times k$ matrix $V_k^T A V_k$, but this is, as we have seen, readily available from the orthogonalization process:

$$V_k^T A V_k = H_{k,k},$$

so that the x_k for which $r_k \perp \mathcal{K}^k(A; r_0)$ can be easily computed by first solving $H_{k,k} y = \|r_0\|_2 e_1$, and then forming $x_k = V_k y$. This algorithm is known as FOM or GENCG [169].

When A is symmetric, then $H_{k,k}$ reduces to a tridiagonal matrix $T_{k,k}$, and the resulting method is known as the *Lanczos* method [130]. When A is in addition positive definite then we obtain, at least formally, the *Conjugate Gradient* method. In commonly used implementations of this method, an LU factorization for $T_{k,k}$ is implicitly formed without generating $T_{k,k}$ itself, and this leads to very elegant short recurrences for the x_j and the corresponding r_j, see Chapter 5.

The positive definiteness is necessary to guarantee the existence of the LU factorization, but it also allows for another useful interpretation. From the fact that $r_i \perp \mathcal{K}^i(A; r_0)$, it follows that $A(x_i - x) \perp \mathcal{K}^i(A; r_0)$, or $x_i - x \perp_A \mathcal{K}^i(A; r_0)$. The latter observation expresses the fact that the error is A-orthogonal to the Krylov subspace and this is equivalent to the important observation that $\|x_i - x\|_A$ is minimal[1]. For an overview of the history of CG and important contributions on this subject see [96].

4.2 The minimum norm residual approach

The creation of an orthogonal basis for the Krylov subspace, with basis vectors $v_1, \ldots, v_{i+1}$, leads to

$$AV_i = V_{i+1}H_{i+1,i}, \tag{4.1}$$

where V_i is the matrix with columns v_1 to v_i. We look for an $x_i \in K^i(A; r_0)$, that is $x_i = V_i y$, for which $\|b - Ax_i\|_2$ is minimal. This norm can be rewritten, with $\rho \equiv \|r_0\|_2$, as

$$\|b - Ax_i\|_2 = \|b - AV_i y\|_2 = \|\rho V_{i+1}e_1 - V_{i+1}H_{i+1,i}y\|_2.$$

Now we exploit the fact that V_{i+1} is an orthonormal transformation with respect to the Krylov subspace $K^{i+1}(A; r_0)$:

$$\|b - Ax_i\|_2 = \|\rho e_1 - H_{i+1,i}y\|_2,$$

and this final norm can simply be minimized by solving the minimum norm least squares problem for the $i+1$ by i matrix $H_{i+1,i}$ and right-hand side $||r_0||_2 e_1$. The least squares problem is solved by constructing a QR factorization of $H_{i+1,i}$, and because of the upper Hessenberg structure that can conveniently be done with Givens transformations [98].

The GMRES method is based upon this approach; see Chapter 6.

4.3 The Petrov–Galerkin approach

For unsymmetric systems we cannot, in general, reduce the matrix A to a symmetric system in a lower-dimensional subspace, by orthogonal projections. The reason is that we cannot create an orthogonal basis for the Krylov subspace

[1] The A-norm is defined by $\|y\|_A^2 \equiv (y, y)_A \equiv (y, Ay)$, and we need the positive definiteness of A in order to get a proper inner product $(\cdot, \cdot)_A$.

by a three-term recurrence relation [79]. We can, however, obtain a suitable non-orthogonal basis with a three-term recurrence, by requiring that this basis be orthogonal with respect to some other basis.

We start by constructing an arbitrary basis for the Krylov subspace:

$$h_{i+1,i}v_{i+1} = Av_i - \sum_{j=1}^{i} h_{j,i}v_j, \tag{4.2}$$

which can be rewritten in matrix notation as $AV_i = V_{i+1}H_{i+1,i}$. The coefficients $h_{i+1,i}$ define the norm of v_{i+1}, and a natural choice would be to select them such that $\|v_{i+1}\|_2 = 1$. In Bi-CG implementations, a popular choice is to select $h_{i+1,i}$ such that $\|v_{i+1}\|_2 = \|r_{i+1}\|_2$.

Clearly, we cannot use V_i for the projection, but suppose we have a W_i for which $W_i^T V_i = D_i$ (an i by i diagonal matrix with diagonal entries d_i), and for which $W_i^T v_{i+1} = 0$. Then

$$W_i^T AV_i = D_i H_{i,i}, \tag{4.3}$$

and now our goal is to find a W_i for which $H_{i,i}$ is tridiagonal. This means that $V_i^T A^T W_i$ should also be tridiagonal. This last expression has a similar structure to the right-hand side in (4.3), with only W_i and V_i reversed. This suggests generating the w_i with A^T.

We choose an arbitrary $w_1 \neq 0$, such that $w_1^T v_1 \neq 0$. Then we generate v_2 with (4.2), and orthogonalize it with respect to w_1, which means that $h_{1,1} = w_1^T Av_1/(w_1^T v_1)$. Since $w_1^T Av_1 = (A^T w_1)^T v_1$, this implies that w_2, generated from

$$h_{2,1}w_2 = A^T w_1 - h_{1,1}w_1,$$

is also orthogonal to v_1.

This can be continued, and we see that we can create bi-orthogonal basis sets $\{v_j\}$ and $\{w_j\}$ by making the new v_i orthogonal to w_1 up to w_{i-1}, and then by generating w_i with the same recurrence coefficients, but with A^T instead of A.

Now we have that $W_i^T AV_i = D_i H_{i,i}$, and also that $V_i^T A^T W_i = D_i H_{i,i}$. This implies that $D_i H_{i,i}$ is symmetric, and hence $H_{i,i}$ is a tridiagonal matrix, which gives us the desired 3-term recurrence relation for the v_js, and the w_js. Note that $v_1, \ldots, v_i$ form a basis for $\mathcal{K}^i(A; v_1)$, and $w_1, \ldots, w_i$ form a basis for $\mathcal{K}^i(A^T; w_1)$.

These bi-orthogonal sets of vectors form the basis for methods such as Bi-CG and QMR.

4.4 The minimum norm error approach

In SYMMLQ [153], with $A = A^T$ we minimize the norm of the error $x - x_k$, for $x_k = x_0 + AV_k y_k$, which means that y_k is the solution of the normal equations

$$V_k^T AAV_k y_k = V_k^T A(x - x_0) = V_k^T r_0 = ||r_0||_2 e_1. \tag{4.4}$$

This system can be further simplified by exploiting the Lanczos relations (3.18), with $T_{k+1,k} \equiv H_{k+1,k}$:

$$V_k^T AAV_k = T_{k+1,k}^T V_{k+1}^T V_{k+1} T_{k+1,k} = T_{k+1,k}^T T_{k+1,k}. \tag{4.5}$$

A stable way of solving this set of normal equations is based on an $L\widetilde{Q}$ decomposition of $T_{k+1,k}^T$, but note that this is equivalent to the transpose of the $Q_{k+1,k}R_k$ decomposition of $T_{k+1,k}$, constructed for MINRES (by Givens rotations), and where R_k is an upper tridiagonal matrix (only the diagonal and the first two co-diagonals in the upper triangular part contain nonzero elements):

$$T_{k+1,k}^T = R_k^T Q_{k+1,k}^T.$$

This leads to

$$T_{k+1,k}^T T_{k+1,k} y_k = R_k^T R_k y_k = ||r_0||_2 e_1,$$

from which the basic generating formula for SYMMLQ is obtained:

$$\begin{aligned} x_k &= x_0 + AV_k R_k^{-1} R_k^{-T} ||r_0||_2 e_1 \\ &= x_0 + V_{k+1} Q_{k+1,k} R_k R_k^{-1} R_k^{-T} ||r_0||_2 e_1 \\ &= x_0 + (V_{k+1} Q_{k+1,k}) \, (L_k^{-1} ||r_0||_2 e_1), \end{aligned} \tag{4.6}$$

with $L_k \equiv R_k^T$. The actual implementation of SYMMLQ [153] is based on an update procedure for $W_k \equiv V_{k+1} Q_{k+1,k}$, and on a three-term recurrence relation for $||r_0||_2 L_k^{-1} e_1$.

In SYMMLQ it is possible to generate the Galerkin approximations as a side product. This means that for positive definite symmetric matrices the CG results can be reconstructed, at relatively low costs, from the SYMMLQ results [153]. This gives no advantage for positive definite matrices, but it can be used for indefinite symmetric matrices. The advantage over CG in that situation is that SYMMLQ avoids having to construct the LU decomposition of $T_{k,k}$ and the latter may not exist (without pivoting), or be singular, for indefinite systems.

5

The Conjugate Gradients method

5.1 Derivation of the method

As explained in Section 4.1, the Conjugate Gradients method can be viewed as a variant of the Lanczos method. The method is based on relation (3.18), which for symmetric A reduces to $AV_i = V_{i+1}H_{i+1,i}$ with tridiagonal $H_{i+1,i}$. For the k-th column of V_k, we have that

$$Av_k = h_{k+1,k}v_{k+1} + h_{k,k}v_k + h_{k-1,k}v_{k-1}. \tag{5.1}$$

In the Galerkin approach, the new residual $b - Ax_k$ is orthogonal to the subspace spanned by $v_1, \ldots, v_k$, so that r_k is in the direction of v_{k+1}. Therefore, we can also select the scaling factor $h_{k+1,k}$ so that v_{k+1} coincides with r_k. This would be convenient, since the residual gives useful information on our solution, and we do not want to work with two sequences of auxiliary vectors.

From the consistency relation (3.12) we have that r_k can be written as

$$r_k = (I - AQ_{k-1}(A))r_0.$$

By inserting the polynomial expressions for the residuals in (5.1), and comparing the coefficient for r_0 in the new relation, we obtain

$$h_{k+1,k} + h_{k,k} + h_{k-1,k} = 0,$$

which defines $h_{k+1,k}$.

At the end of this section we will consider the situation where the recurrence relation terminates.

With R_i we denote the matrix with columns r_j:

$$R_i = (r_0, \ldots, r_{i-1}),$$

then we have

$$AR_i = R_{i+1}T_{i+1,i}, \tag{5.2}$$

where $T_{i+1,i}$ is a tridiagonal matrix (with $i+1$ rows and i columns); its nonzero elements are defined by the $h_{i,j}$.

Since we are looking for a solution x_i in $\mathcal{K}^i(A; r_0)$, that vector can be written as a combination of the basis vectors of the Krylov subspace, and hence

$$x_i = R_i y.$$

Note that y has i components.

Furthermore, the Ritz–Galerkin condition says that the residual for x_i is orthogonal with respect to $r_0, \ldots, r_{i-1}$:

$$R_i^T(Ax_i - b) = 0,$$

and hence

$$R_i^T AR_i y - R_i^T b = 0.$$

Using equation (5.2), we obtain

$$R_i^T R_i T_{i,i} y = \|r_0\|_2^2 e_1$$

Since $R_i^T R_i$ is a diagonal matrix with diagonal elements $\|r_0\|_2^2$ up to $\|r_{i-1}\|_2^2$, we find the desired solution by solving y from

$$T_{i,i} y = e_1 \quad \Rightarrow \quad y \quad \Rightarrow \quad x_i = R_i y.$$

So far we have only used the fact that A is symmetric and we have assumed that the matrix T_i is not singular. The Krylov subspace method that has been derived here is known as the Lanczos method for symmetric systems [130]. We will exploit the relation between the Lanczos method and the conjugate gradients method for the analysis of the convergence behaviour of the latter method.

Note that for some $j \leq n-1$ the construction of the orthogonal basis must terminate. In that case we have that $AR_{j+1} = R_{j+1}T_{j+1,j+1}$. Let y be the solution of the reduced system $T_{j+1,j+1}y = e_1$, and $x_{j+1} = R_{j+1}y$. Then it follows that $x_{j+1} = x$, i.e., we have arrived at the exact solution, since $Ax_{j+1} - b = AR_{j+1}y - b = R_{j+1}T_{j+1,j+1}y - b = R_{j+1}e_1 - b = 0$ (we have assumed that $x_0 = 0$).

Exercise 5.1. *Show that exact termination for a* $j \leq n-1$ *also holds if* $x_0 \neq 0$.

The Conjugate Gradients method [115], CG for short, is a clever variant on the above approach and saves storage and computational effort. If we follow naively the above sketched approach when solving the projected equations, then we see that we have to save all columns of R_i throughout the process in order to recover the current iteration vectors x_i. This can be done in a more memory friendly way. If we assume that the matrix A is in addition positive definite then, because of the relation

$$R_i^T A R_i = R_i^T R_i T_{i,i},$$

we conclude that $T_{i,i}$ can be transformed by a rowscaling matrix $R_i^T R_i$ into a positive definite symmetric tridiagonal matrix (note that $R_i^T A R_i$ is positive definite for $y \in \mathbb{R}^i$). This implies that $T_{i,i}$ can be LU decomposed without any pivoting:

$$T_{i,i} = L_i U_i,$$

with L_i lower bidiagonal, and U_i upper bidiagonal with unit diagonal. Hence

$$x_i = R_i y = R_i T_{i,i}^{-1} e_1 = (R_i U_i^{-1})(L_i^{-1} e_1). \tag{5.3}$$

We concentrate on the factors, placed between parentheses, separately.

(1)

$$L_i = \begin{pmatrix} \delta_0 & & & & \\ \phi_0 & \delta_1 & & & \\ & \phi_1 & \ddots & & \\ & & \ddots & \ddots & \\ & & & \phi_{i-2} & \delta_{i-1} \end{pmatrix}$$

With $q \equiv L_i^{-1} e_1$ we have that q can be solved from $L_i q = e_1$. Hence $q_0 = 1/\delta_0$, and $q_{i-1} = -\phi_{i-2} q_{i-2} / \delta_{i-1}$, so that the elements of q can be computed in a recursive manner.

(2) Defining $P_i \equiv R_i U_i^{-1}$, we have that

$$R_i = P_i U_i = \begin{pmatrix} p_0 & \cdots & p_{i-2} & p_{i-1} \end{pmatrix} \begin{pmatrix} 1 & \epsilon_0 & & \\ & 1 & \epsilon_1 & \ddots \\ & & \ddots & \epsilon_{i-2} \\ & & & 1 \end{pmatrix}$$

$$\Rightarrow \quad r_{i-1} = \epsilon_{i-2} p_{i-2} + p_{i-1},$$

so that the vector p_{i-1} can be recursively computed as

$$p_{i-1} = r_{i-1} - \epsilon_{i-2} p_{i-2}.$$

Exercise 5.2. *Show that in step i from this process we can simply expand L_i to L_{i+1}, and U_i to U_{i+1}. Hence, we can simply compute q_i and p_i, using the above obtained recursion relations with the new values ϕ_{i-1}, δ_i, and ϵ_{i-1}.*

Pasting the two recurrences together we obtain

$$\begin{aligned} x_i &= \begin{pmatrix} p_0 & \cdots & p_{i-1} \end{pmatrix} \begin{pmatrix} \vdots \\ \vdots \\ q_{i-1} \end{pmatrix} \\ &= x_{i-1} + q_{i-1} p_{i-1}. \end{aligned}$$

In principle the method is not too complicated: the tridiagonal matrix is generated from a simple three-term recurrence, and this matrix is factorized and solved for both factors. However, as we will see, *it is not necessary to generate $T_{i,i}$ explicitly*, we can obtain the required information in an easier way.

To see this, we simplify the notation for our recurrence relations and then we exploit the orthogonality properties of the underlying Lanczos method. First we write $\alpha_i \equiv q_i$, and $\beta_i \equiv e_i$.

Then our two-term recurrence relations can be recast as

$$p_{i-1} = r_{i-1} + \beta_{i-2} p_{i-2} \tag{5.4}$$

$$x_i = x_{i-1} + \alpha_{i-1} p_{i-1} \tag{5.5}$$

$$r_i = r_{i-1} - \alpha_{i-1} A p_{i-1}. \tag{5.6}$$

Exercise 5.3. *Show that $p_i^T A p_j = 0$ for $i \neq j$.*

The vector $\alpha_{i-1} p_{i-1}$ is the correction vector that leads to the new minimum of $\|x - x_i\|_A$. It is thus tangential to the surface $\|x - z\|_A = \|x - x_i\|_A$, for $z \in \mathcal{K}^{i+1}(A; r_0)$. The vectors p_j are A-orthogonal, and can be interpreted as *conjugate* (= A-orthogonal) *gradients* for $\|x - z\|_A$, as a function of z. This gave the method its name.

From $r_i^T r_{i-1} = 0$, we derive α_{i-1}:

$$\alpha_{i-1} = \frac{r_{i-1}^T r_{i-1}}{r_{i-1}^T A p_{i-1}},$$

and using relation (5.4), we arrive at the elegant expression

$$\alpha_{i-1} = \frac{r_{i-1}^T r_{i-1}}{p_{i-1}^T A p_{i-1}}.$$

Exercise 5.4. *Show that* $r_{i-1}^T A p_{i-1} = p_{i-1}^T A p_{i-1}$.

For β_{i-1}, we can derive a similar elegant expression. First we multiply the recursion (5.4) for p_{i-1} by A:

$$Ap_{i-1} = Ar_{i-1} + \beta_{i-2} A p_{i-2},$$

and we eliminate Ap_{i-2} with the recurrence relation (5.6), which leads to

$$r_i = r_{i-1} - \alpha_{i-1} A r_{i-1} - \frac{\alpha_{i-1}\beta_{i-2}}{\alpha_{i-2}}(r_{i-2} - r_{i-1}). \tag{5.7}$$

Since $r_i^T r_{i-2} = 0$, we obtain

$$\beta_{i-2} = -\alpha_{i-2} \frac{r_{i-2}^T A r_{i-1}}{r_{i-2}^T r_{i-2}} = -\alpha_{i-2} \frac{r_{i-1}^T A r_{i-2}}{r_{i-2}^T r_{i-2}}$$
$$= \frac{r_{i-1}^T r_{i-1}}{r_{i-2}^T r_{i-2}} \quad \text{(cf. (5.7))}.$$

Note that we need only two new inner products in iteration step i for the computation of the two iteration coefficients (precisely as many as for the Lanczos process).

Thus we have arrived at the well-known conjugate gradients method. The name stems from the property that the update vectors p_i are A-orthogonal. Note that the positive definiteness of A is exploited only to guarantee safe decomposition of the implicitly generated tridiagonal matrix $T_{i,i}$. This suggests that the conjugate gradients method may also work for certain nonpositive definite systems, but then at our own risk [152]. We will see later how other ways of solving the projected system lead to other well-known methods.

5.2 Computational notes

The standard (unpreconditioned) Conjugate Gradient method for the solution of $Ax = b$ can be represented by the scheme in Figure 5.1.

x_0 is an initial guess, $r_0 = b - Ax_0$
for $i = 1, 2, \ldots$.
  $\rho_{i-1} = r_{i-1}^T r_{i-1}$
  if $i = 1$
    $p_i = r_{i-1}$
  else
    $\beta_{i-1} = \rho_{i-1}/\rho_{i-2}$
    $p_i = r_{i-1} + \beta_{i-1} p_{i-1}$
  endif
  $q_i = Ap_i$
  $\alpha_i = \rho_{i-1}/p_i^T q_i$
  $x_i = x_{i-1} + \alpha_i p_i$
  $r_i = r_{i-1} - \alpha_i q_i$
  if x_i accurate enough **then** quit
end

Figure 5.1. Conjugate Gradients without preconditioning.

Exercise 5.5. *Consider the Conjugate Gradients scheme in Figure 5.1 for a given linear system $Ax = b$, with starting vector x_0. Now consider the application of this CG scheme for the system*

$$\widetilde{A}\widetilde{x} = \widetilde{b}, \tag{5.8}$$

with $\widetilde{A} = Q^T AQ$, $\widetilde{x} = Q^T x$, and $\widetilde{b} = Q^T b$, and Q an orthonormal matrix: $Q^T Q = I$. Denote the computed variables for CG applied to (5.8) with a superscript $\widetilde{}$, and start the iteration with $\widetilde{x}_0 = Q^T x_0$.

Show that the scheme applied for (5.8) generates the same iteration constants as in the iteration process for $Ax = b$. Show also that $\|\widetilde{r}_i\|_2 = \|r_i\|_2$.
Obviously, an orthonormal transformed system leads to the same CG iteration process.

Exercise 5.6. *Show that for studying the convergence behaviour of CG it is no restriction if we restrict ourselves to diagonal matrices A (except for rounding errors). Verify this by numerical experiments. Hint: use the results of the previous exercise.*

Exercise 5.7. *Do experiments with CG for suitably chosen $Ax = b$, and for the system that is orthogonally transformed with Q so that $Q^T AQ$ is a diagonal* *matrix. Any difference between the convergence histories must be attributed to rounding errors. Is there much difference? Enough difference to prevent us from drawing conclusions on the convergence behaviour of CG if we do experiments with diagonal matrices?*

Exercise 5.8. *Do experiments with CG for diagonal matrices with only k different eigenvalues. What is the maximum number of iterations to obtain a reduction in the norm of the residual by, say,* 10^{-10}*? Explain this result. Hint: use the polynomial interpretation for the residuals in CG.*

CG is most often used in combination with a suitable approximation K for A and then K is called the preconditioner. We will assume that K is also positive definite. However, we cannot apply CG straightaway for the explicitly preconditioned system $K^{-1}Ax = K^{-1}b$, as we suggested in the introduction, because $K^{-1}A$ is most likely not symmetric. One way out is to apply the preconditioner differently. Assume that K is given in factored form:

$$K = LL^T,$$

as is the case for ILU preconditioners.

We then apply CG for the symmetrically preconditioned system

$$L^{-1}AL^{-T}y = L^{-1}b,$$

with $x = L^{-T}y$.

This approach has the disadvantage that K must be available in factored form and that we have to backtransform the approximate solution afterwards. There is a more elegant alternative. Note first that the CG method can be derived for any choice of the inner product. In our derivation we have used the standard inner product $(x, y) = \sum x_i y_i$, but we have not used any specific property of that inner product. Now we make a different choice:

$$[x, y] \equiv (x, Ky).$$

Exercise 5.9. *Show that* $[x, y]$ *defines a proper inner product.*

It is easy to verify that $K^{-1}A$ is symmetric positive definite with respect to [,]:

$$\begin{aligned}[K^{-1}Ax, y] &= (K^{-1}Ax, Ky) = (Ax, y)\\ &= (x, Ay) = [x, K^{-1}Ay]. \end{aligned} \tag{5.9}$$

x_0 is an initial guess, $r_0 = b - Ax_0$
for $i = 1, 2, \ldots.$
  Solve $Kw_{i-1} = r_{i-1}$
  $\rho_{i-1} = r_{i-1}^H w_{i-1}$
  if $i = 1$
    $p_i = w_{i-1}$
  else
    $\beta_{i-1} = \rho_{i-1}/\rho_{i-2}$
    $p_i = w_{i-1} + \beta_{i-1} p_{i-1}$
  endif
  $q_i = Ap_i$
  $\alpha_i = \rho_{i-1}/p_i^H q_i$
  $x_i = x_{i-1} + \alpha_i p_i$
  $r_i = r_{i-1} - \alpha_i q_i$
  if x_i accurate enough **then** quit
end

Figure 5.2. Conjugate Gradients with preconditioning K.

Hence, we can follow our CG procedure for solving the preconditioned system $K^{-1}Ax = K^{-1}b$, using the new [,]-inner product.

Apparently, we now are minimizing

$$[x_i - x, K^{-1}A(x_i - x)] = (x_i - x, A(x_i - x)),$$

which leads to the remarkable (and known) result that for this preconditioned system we still minimize the error in A-norm, but now over a Krylov subspace generated by $K^{-1}r_0$ and $K^{-1}A$.

In the computational scheme for preconditioned CG, displayed in Figure 5.2, for the solution of $Ax = b$ with preconditioner K, we have again replaced the [,]-inner product by the familiar standard inner product. For example, note that with $\widetilde{r}_{i+1} = K^{-1}Ax_{i+1} - K^{-1}b$ we have that

$$\begin{aligned}\rho_{i+1} &= [\widetilde{r}_{i+1}, \widetilde{r}_{i+1}] \\ &= [K^{-1}r_{i+1}, K^{-1}r_{i+1}] = [r_{i+1}, K^{-2}r_{i+1}] \\ &= (r_{i+1}, K^{-1}r_{i+1}),\end{aligned}$$

and $K^{-1}r_{i+1}$ is the residual corresponding to the preconditioned system $K^{-1}Ax = K^{-1}b$. Furthermore, note that the whole derivation of this method also holds for complex Hermitian matrices A and K, provided we use the standard complex inner product $v^H w$ for vectors v and w in $\mathbb{C}^n$.

Exercise 5.10. *In view of the fact that diagonal matrices can be used to study the convergence behaviour of CG: is it really necessary to construct preconditioners in order to study the effect of clustering of eigenvalues?*

Do experiments with matrices that have clustered eigenvalues and compare the convergence histories with systems that have more uniformly distributed eigenvalue distributions. Does clustering make a difference?

Exercise 5.11. *For some classes of linear systems, particular preconditioners have the effect that they reduce the condition number by a factor, say 4, and that they cluster most of the eigenvalues around 1. Do experiments with diagonal matrices and study the effects.*

If preconditioning would make each iteration step twice as expensive, is preconditioning then likely to lead to efficient iteration processes? Do these experiments prove that preconditioning will be effective? What do they prove?

The coefficients α_j and β_j, generated by the Conjugate Gradient algorithms, as in Figures 5.1 and 5.2, can be used to build the matrix $T_{i,i}$ in the following way:

$$T_{i,i} = \begin{pmatrix} \ddots & & & \\ \ddots & -\frac{\beta_{j-1}}{\alpha_{j-1}} & & \\ \ddots & \frac{1}{\alpha_j} + \frac{\beta_{j-1}}{\alpha_{j-1}} & \ddots & \\ & -\frac{1}{\alpha_j} & \ddots & \\ & & & \ddots \end{pmatrix}. \tag{5.10}$$

With the matrix $T_{i,i}$ approximations can easily be computed for eigenvalues of A, the so-called *Ritz values*. To this end, note that

$$R_i^T A R_i = D_i T_{i,i},$$

where D_i is a diagonal matrix with diagonal elements $\|r_0\|_2^2$, up to $\|r_i\|_2^2$. Multiplication from the left and the right by the matrix $D_i^{-\frac{1}{2}}$ leads to

$$D_i^{-\frac{1}{2}} R_i^T A R_i D_i^{-\frac{1}{2}} = D_i^{\frac{1}{2}} T_{i,i} D_i^{-\frac{1}{2}}.$$

Exercise 5.12. *Show that the n by $i+1$ matrix $R_i D_i^{-\frac{1}{2}}$ is an orthonormal matrix, that is $(R_i D_i^{-\frac{1}{2}})^T R_i D_i^{-\frac{1}{2}} = I_{i+1,i+1}$.*

Apparently, the $i+1$ columns of $R_i D_i^{-\frac{1}{2}}$ span an orthonormal basis for the subspace $\mathcal{K}^{i+1}(A; r_0)$. A vector x from that subspace can be expressed as

$$x = R_i D_i^{-\frac{1}{2}} y.$$

If we take Rayleigh quotients for A with vectors $x \neq 0$ from the Krylov subspace, we have

$$\begin{aligned}\rho(A, x) &\equiv \frac{(Ax, x)}{(x, x)} \\ &= \frac{(A R_i D_i^{-\frac{1}{2}} y, R_i D_i^{-\frac{1}{2}} y)}{(y, y)} \\ &= \frac{(D_i^{\frac{1}{2}} T_{i,i} D_i^{-\frac{1}{2}} y, y)}{(y, y)}.\end{aligned}$$

The eigenvalues of the matrix $D_i^{\frac{1}{2}} T_{i,i} D_i^{-\frac{1}{2}}$ represent the local extrema of the Rayleigh quotient for A with respect to the Krylov subspace. We denote the eigenpairs of $T_{i,i}$ by

$$\theta_j^{(i)}, z_j^{(i)}.$$

Exercise 5.13. *Show that if $D_i^{\frac{1}{2}} T_{i,i} D_i^{-\frac{1}{2}} z_j^{(i)} = \theta_j^{(i)} z_j^{(i)}$, then*

$$A R_i D_i^{-\frac{1}{2}} z_j^{(i)} - \theta_j^{(i)} R_i D_i^{-\frac{1}{2}} z_j^{(i)} \perp \mathcal{K}^{i+1}(A; r_0).$$

We conclude that the residual for the approximate eigenpair

$$\theta_j^{(i)}, x_j^{(i)} \equiv R_i D_i^{-\frac{1}{2}} z_j^{(i)}$$

is orthogonal to the current Krylov subspace. For this reason, this pair is usually referred to as a (Rayleigh–)Ritz value and its corresponding Ritz vector.

Exercise 5.14. *Show that the matrix* $\widetilde{T}_{i,i} \equiv D_i^{\frac{1}{2}} T_{i,i} D_i^{\frac{1}{2}}$ *is a symmetric matrix that can be expressed as the following symmetric tridiagonal matrix:*

$$\widetilde{T}_i = \begin{pmatrix} \ddots & & & \\ \ddots & -\frac{\sqrt{\beta_{j-1}}}{\alpha_{j-1}} & & \\ \ddots & \frac{1}{\alpha_j} + \frac{\beta_{j-1}}{\alpha_{j-1}} & \ddots & \\ & -\frac{\sqrt{\beta_j}}{\alpha_j} & \ddots & \\ & & \ddots & \end{pmatrix}.$$

The eigenvalues of the leading i-th order minor of this matrix are the Ritz values of A (for Figure 5.1) or the preconditioned matrix $K^{-1}A$ (for Figure 5.2) with respect to the i-dimensional Krylov subspace spanned by the first i residual vectors. The Ritz values approximate the (extremal) eigenvalues of the (preconditioned) matrix increasingly well. These approximations can be used to get an impression of the relevant eigenvalues. They can also be used to construct upperbounds for the error in the delivered approximation with respect to the solution [124, 113]. According to the results in [191] the eigenvalue information can also be used in order to understand or explain delays in the convergence behaviour.

Exercise 5.15. *Do experiments with CG for several linear systems with diagonal matrices. Compute the Ritz values for* T_i *for some values of* i. *What observations can be made? Can some relation between the Ritz values and convergence of CG be discerned?*

5.3 The convergence of Conjugate Gradients

Exercise 5.16. *Show that* $x - u = Q_i(A)(x - x_0)$, *for some polynomial* Q_i *of degree* i *with* $Q_i(0) = 1$, *if* $u \in x_0 + \mathcal{K}^i(A; r_0)$.

The conjugate gradient method (here with $K = I$) constructs in the i-th iteration step an x_i, that can be written as

$$x_i - x = P_i(A)(x_0 - x), \tag{5.11}$$

such that $\|x_i - x\|_A$ is minimal over all polynomials P_i of degree i, with $P_i(0) = 1$. Let us denote the eigenvalues and the orthonormalized eigenvectors of A by λ_j, z_j. We write $r_0 = \sum_j \gamma_j z_j$. It follows that

$$r_i = P_i(A) r_0 = \sum_j \gamma_j P_i(\lambda_j) z_j \tag{5.12}$$

and hence

$$\|x_i - x\|_A^2 = \sum_j \frac{\gamma_j^2}{\lambda_j} P_i^2(\lambda_j). \tag{5.13}$$

Note that only those λ_j play a role in this process for which $\gamma_j \neq 0$. In particular, if A happens to be semidefinite, i.e., there is a $\lambda = 0$, then this is no problem for the minimization process as long as the corresponding coefficient γ is zero as well. The situation where γ is small, due to rounding errors, is discussed in [124].

Exercise 5.17. *Write $b = \sum_{j=1}^{n} \gamma_j z_j$. Consider only those values of j for which $\gamma_j \neq 0$. Let ℓ denote the number of* different *eigenvalues associated with these nonzero γs. Show that CG delivers the exact solution in ℓ iteration steps. Hint: use the polynomial expression for the iterates.*

Upperbounds on the error (in A-norm) are obtained by observing that

$$\begin{aligned} \|x_i - x\|_A^2 &= \sum_j \frac{\gamma_j^2}{\lambda_j} P_i^2(\lambda_j) \leq \sum_j \frac{\gamma_j^2}{\lambda_j} Q_i^2(\lambda_j) \\ &\leq \max_{\lambda_j} Q_i^2(\lambda_j) \sum_j \frac{\gamma_j^2}{\lambda_j}, \end{aligned} \tag{5.14}$$

for any arbitrary polynomial Q_i of degree i with $Q_i(0) = 1$, where the maximum is taken, of course, only over those λ for which the corresponding $\gamma \neq 0$.

When P_i has zeros at all the different λ_j then $r_i = 0$. The conjugate gradients method tries to spread the zeros in such a way that $P_i(\lambda_j)$ is small in a weighted sense, i.e., $\|x_i - x\|_A$ is as small as possible.

We get descriptive upperbounds by selecting appropriate polynomials for Q_i. A very well-known upperbound arises by taking for Q_i the i-th degree Chebyshev polynomial C_i transformed to the interval $[\lambda_{min}, \lambda_{max}]$ and scaled such that its value at 0 is equal to 1:

$$Q_i(\lambda) = \frac{C_i\left(\frac{2\lambda - (\lambda_{min} + \lambda_{max})}{\lambda_{max} - \lambda_{min}}\right)}{C_i\left(-\frac{\lambda_{min} + \lambda_{max}}{\lambda_{max} - \lambda_{min}}\right)}.$$

For the Chebyshev polynomials C_i we have the following properties:

$$C_i(x) = \cos(i \ \arccos(x)) \ \text{for} \ -1 \le x \le 1, \tag{5.15}$$

$$|C_i(x)| \le 1 \ \text{for} \ -1 \le x \le 1, \tag{5.16}$$

$$C_i(x) = \frac{1}{2}\left[(x+\sqrt{x^2-1})^i + (x+\sqrt{x^2-1})^{-i}\right] \ \text{for} \ x \ge 1, \tag{5.17}$$

$$|C_i(x)| = |C_i(-x)|. \tag{5.18}$$

With Q_i instead of the optimal CG-polynomial P_i, we have that

$$\|x_i - x\|_A^2 \le \sum \frac{\gamma_j^2}{\lambda_j} Q_i^2(\lambda_j)$$
$$\le \max_{\lambda_{min},\lambda_{max}} |Q_i^2(\lambda_j)| \, \|x_0 - x\|_A^2. \tag{5.19}$$

Using properties of the Chebyshev polynomials, we can now derive an elegant upperbound for the error in A-norm . First we note that

$$|Q_i(\lambda)| \le \frac{1}{|C_i(\frac{\lambda_{max}+\lambda_{min}}{\lambda_{max}-\lambda_{min}})|} \quad \text{for } \lambda_{min} \le \lambda \le \lambda_{max}.$$

Exercise 5.18. *Show that for* $0 < a < b$ *we have that*

$$C_i(\frac{b+a}{b-a}) \ge \frac{1}{2}(x+\sqrt{x^2-1})^i,$$

with $x = (b+a)/(b-a)$. *Note that this lower bound is increasingly sharp for increasing* i.

With $a \equiv \lambda_{min}$ and $b \equiv \lambda_{max}$, we have for $a \le \lambda \le b$ that

$$|Q_i\lambda| = \frac{1}{|C_i(\frac{b+a}{b-a})|} \tag{5.20}$$

$$\le 2\left(\frac{1}{x+\sqrt{x^2-1}}\right)^i \quad \text{with } x \equiv \frac{b+a}{b-a} \tag{5.21}$$

$$= \left(x - \sqrt{x^2-1}\right)^i \tag{5.22}$$

$$= \left(\frac{b+a-2\sqrt{ba}}{b-a}\right)^i \tag{5.23}$$

$$= \left(\frac{\sqrt{b}-\sqrt{a}}{\sqrt{b}+\sqrt{a}}\right)^i. \tag{5.24}$$

With $\kappa \equiv b/a = \lambda_{max}/\lambda_{min}$, we obtain a well known upperbound for the A-norm of the error [44, 98, 8]:

$$\|x_i - x\|_A \leq 2\left(\frac{\sqrt{\kappa}-1}{\sqrt{\kappa}+1}\right)^i \|x_0 - x\|_A. \tag{5.25}$$

This upperbound shows that we have fast convergence for small condition numbers. It was shown by Axelsson [8] that similar upperbounds can be obtained for specific eigenvalue distributions. For instance, let us consider the situation that $\lambda_n > \lambda_{n-1}$. Then we take for Q_i, the polynomial

$$Q_i(\lambda) = \frac{\lambda_n - \lambda}{\lambda_n} C_{i-1}\left(\frac{2\lambda - (a+b)}{b-a}\right) / C_{i-1}\left(\frac{-(a+b)}{b-a}\right),$$

with $a \equiv \lambda_{min}$, $b \equiv \lambda_{n-1}$.

This shows that for the situation where λ_{n-1} is relatively much smaller than λ_n, the upperbound for the error for the CG process lags only one step behind the upperbound for a process with a condition number λ_{n-1}/λ_1.

Exercise 5.19. *Derive an upperbound for the A-norm of the error that shows that this error is at most two steps behind the upperbound for a process with condition number* λ_{n-2}/λ_1.

These types of upperbounds show that it is important to have small condition numbers, or, in the case of larger condition numbers, to have eigenvalue distributions with well-separated eigenvalues that cause these large condition numbers. In that case we say that the (remaining) part of the spectrum is (relatively) clustered. The purpose of preconditioning is to reduce the condition number κ and/or clustering of the eigenvalues.

In [8] the situation is analysed where the eigenvalues of $K^{-1}A$ are in disjoint intervals.

5.3.1 Local effects in the convergence behaviour

Upperbounds as in (5.25) show that we have global convergence, but they do not help us to explain all sorts of local effects in the convergence behaviour of CG. A very well-known effect is the so-called superlinear convergence: in many situations the average speed of convergence seems to increase as the iteration proceeds. As we have seen, the conjugate gradients algorithm is just an efficient implementation of the Lanczos algorithm. The eigenvalues of the implicitly generated tridiagonal matrix T_i are the Ritz values of A with respect to the current Krylov subspace. It is known from Lanczos theory that these Ritz values converge towards the eigenvalues of A and that in general the extreme eigenvalues of A are well approximated during early iterations [129, 149, 155]. Furthermore, the speed of convergence depends on how well these eigenvalues are separated from the others (gap ratio) [155]. This will help us to understand the superlinear convergence behaviour of the conjugate gradient method (as well as other Krylov subspace methods). It can be shown that as soon as one of the extreme eigenvalues is modestly well approximated by a Ritz value, the procedure converges from then on as a process in which this eigenvalue is absent, i.e., a process with a reduced condition number. Note that superlinear convergence behaviour in this connection is used to indicate linear convergence with a factor that is gradually decreased during the process as more and more of the extreme eigenvalues are sufficiently well approximated.

Example We consider a linear system of order $n = 100$ with one isolated eigenvalue 0.0005 and the other 99 eigenvalues equally distributed over the interval [0.08, 1.21]. The right-hand side was chosen so that the initial residual has equal components in all eigenvector directions.

Exercise 5.20. *Show that this test example can be realized, without loss of generality, by taking a diagonal matrix for A, with the eigenvalues on the diagonal, $x_0 = 0$, and with right-hand side b being the vector with all elements equal to* 1.

The A-norms of the error $x - x_i$ are plotted on a logarithmic scale in Figure 5.3. We see, in this figure, that the convergence for the first eight or so iteration steps is relatively slow, in agreement with the condition number of A: 242.0. After iteration 10 the convergence is noticeably faster: about four times faster than in the earlier phase. This corresponds to a condition number of about 15.1 and that is the condition number for a matrix with eigenvalue spectrum [0.08, 1.21], that is, the smallest eigenvalue 0.005 seems to have lost its influence on the convergence behaviour after 10 iteration steps. The smallest Ritz value at iteration step 10 is $\theta_1^{(10)} \approx 0.0061$, at iteration step 11: $\theta_1^{(11)} \approx 0.0054$, and at step 12: $\theta_1^{(12)} \approx 0.0052$. Clearly, the smallest Ritz value is converging towards the smallest eigenvalue $\lambda_1 = 0.005$, but we see that the faster convergence of CG commences already in a phase where the smallest Ritz value is still a relatively crude approximation (at step 10 we have an error of about 20%). We will now analyse this behaviour in more detail.

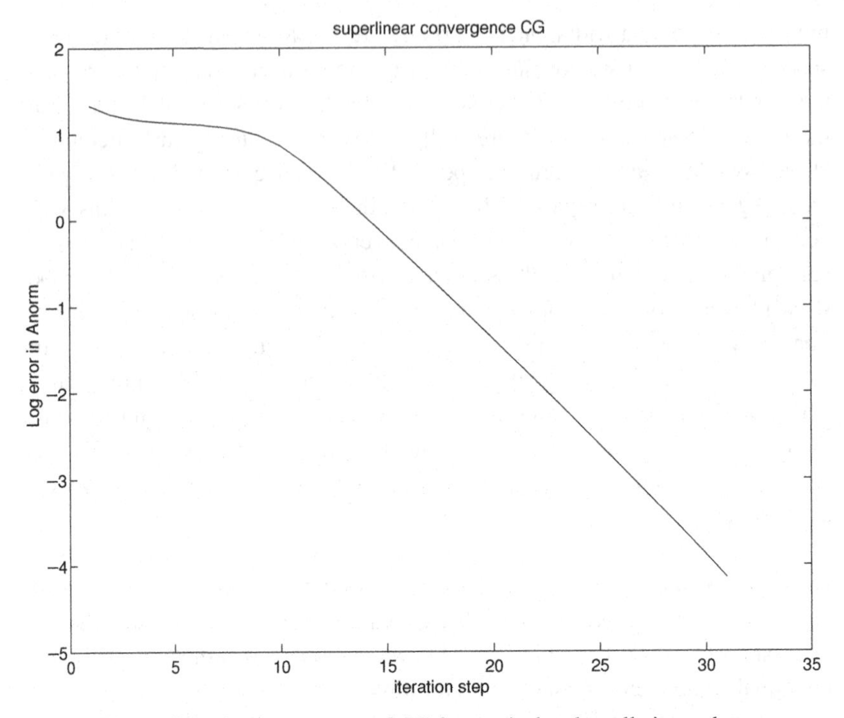

Figure 5.3. The convergence of CG for one isolated small eigenvalue.

The local convergence behaviour of CG, and especially the occurrence of superlinear convergence, was first explained in a qualitative sense in [44], and later in a quantitative sense in [191]. In both papers it was linked to the convergence of eigenvalues (Ritz values) of $T_{i,i}$ towards eigenvalues of $K^{-1}A$, for increasing i. Here we will follow the exposition given in [191]. The main idea in that paper is to link the convergence of CG in a certain phase with a process in which one or more of the Ritz values are replaced by eigenvalues of the matrix. To that end we first have to make a link with the Ritz values.

As we have seen, the residual r_i can be expressed as

$$r_i = P_i(A)b, \tag{5.26}$$

with $r_i \perp \mathcal{K}^i(A, r_0)$. Let the orthonormal Lanczos basis for this Krylov subspace be given by $v_1, v_2, \ldots, v_i$. We have seen that r_i is a multiple of v_{i+1}. In the generic case, $v_{i+1} \neq 0$ (if the equal sign holds then we have exact termination), the polynomial P_i has exact degree i. We will now show that the Ritz values are zeros of P_i. We start with the basic relation (3.18), which is rewritten for the symmetric case as

$$AV_i = V_i T_{i,i} + t_{i+1,i} v_{i+1} e_i^T, \tag{5.27}$$

with $T_{i,i}$ a symmetric tridiagonal matrix. After post-multiplication with e_1 (the first canonical unit vector of length i) we obtain

$$Av_1 = V_i T_{i,i} e_1.$$

Pre-multiplying this with A leads to

$$\begin{aligned} A^2 v_1 &= A V_i T_{i,i} e_1 \\ &= (V_i T_{i,i} + t_{i+1,i} v_{i+1} e_i^T) T_{i,i} e_1 \\ &= V_i T_{i,i}^2 e_1 \;\text{ if }\; i > 2 \end{aligned}$$

Exercise 5.21. *Show that*

$$A^i v_1 = V_i T_{i,i} e_1 + c v_{i+1}, \tag{5.28}$$

for some constant c.

Exercise 5.22. *Show that*

$$V_i^T P_i(A) v_1 = P_i(T_{i,i}) e_1 = 0, \tag{5.29}$$

with P_i the polynomial of (5.26).

Equation (5.29) shows that $P_i(T_{i,i})$ is at least singular. We will now show that $P_i(T_{i,i})$ is identical to zero for all eigenvectors of $T_{i,i}$.

The Ritz pairs θ, z of A with respect to $\mathcal{K}^i(A; v_1)$ are defined by

$$Az - \theta z \perp \mathcal{K}^i(A; v_1),$$

for $z \in \mathcal{K}^i(A; v_1)$. With $z = V_i y$ it follows that

$$V_i^T A V_i y - \theta y = 0,$$

so that the pair θ, y is an eigenpair of $T_{i,i}$. Let the eigenpairs of $T_{i,i}$ be denoted as

$$\theta_j^{(i)}, y_j^{(i)}.$$

Since $T_{i,i}$ is symmetric the normalized eigenvectors span a complete orthonormal basis for $\mathbb{R}^i$. This implies that we can express e_1 in terms of this eigenvector basis

$$e_1 = \sum_j \gamma_j y_j^{(i)}. \tag{5.30}$$

Assume that $\gamma_j = (y_j^{(i)}, e_1) = 0$, then because

$$T_{i,i} y_j^{(i)} = \theta_j^{(i)} y_j^{(i)},$$

we have that $(T_{i,i} y_j^{(i)}, e_1) = (y_j^{(i)}, T_{i,i} e_1) = 0$ and this implies that $(y_j^{(i)}, e_2) = 0$. It is now easy to prove that $y_j^{(i)} \perp \{e_1, e_2, \ldots, e_i\}$ and this leads to a contradiction.

A combination of (5.29) with (5.30) gives

$$\sum_j \gamma_j P_i(\theta_j^{(i)}) y_j^{(i)} = 0,$$

and since all γ_j are nonzero it follows that the i Ritz values $\theta_j^{(i)}$ are the zeros of P_i. Because $P_i(0) = 1$ (cf. (3.12)), this polynomial can be fully characterized as

$$P_i(t) = \frac{(\theta_1^{(i)} - t)(\theta_2^{(i)} - t) \cdots (\theta_i^{(i)} - t)}{\theta_1^{(i)} \theta_2^{(i)} \cdots \theta_i^{(i)}}. \tag{5.31}$$

We can now further analyse the convergence behaviour of CG by exploiting the characterization of the residual polynomial P_i. Let λ_j, z_j denote the eigenpairs of A, with normalized z_j. We write the initial error $x - x_0$ as

$$x - x_0 = \sum_1^n \mu_j z_j, \tag{5.32}$$

and this implies, on account of (5.11),

$$x - x_i = \sum_1^n \mu_j P_i(\lambda_j) z_j.$$

Now we replace the first component of x_i by the first component of x and we take the modified x_i as the starting vector $\bar{x}_0$ for another CG process (with the same A and b):

$$x - \bar{x}_0 = \sum_2^n \mu_j P_i(\lambda_j) z_j.$$

This new CG process generates $\bar{x}_k$, for which the error is characterized by a polynomial $\bar{P}_k$. Because of the optimality property of these polynomials, we may replace any of the residual polynomials by arbitrary other polynomials in order to derive upperbounds. We select

$$q_i(t) = \frac{(\lambda_1 - t)(\theta_2^{(i)} - t)\cdots(\theta_i^{(i)} - t)}{\lambda_1 \theta_2^{(i)} \cdots \theta_i^{(i)}} = \frac{\theta_1^{(i)}(\lambda_1 - t)}{\lambda_1(\theta_1^{(i)} - t)} P_i(t).$$

The polynomial $\bar{P}_k q_i$ takes the value 1 for $t = 0$, and it follows that

$$\|x - x_{i+k}\|_A^2 \le \|\bar{P}_k(A) q_i(A)(x - x_0)\|_A^2 \tag{5.33}$$

$$= \sum_2^n \lambda_j \bar{P}_k(\lambda_j)^2 q_i(\lambda_j)^2 \mu_j^2. \tag{5.34}$$

Defining

$$F_i \equiv \frac{\theta_1^{(i)}}{\lambda_1} \max_{j \ge 2} \left| \frac{\lambda_j - \lambda_1}{\lambda_j - \theta_1^{(i)}} \right|,$$

it follows that

$$|q_i(\lambda_j)| \le F_i |P_i(\lambda_j)| \quad \text{for} \quad j \ge 2.$$

We can now further simplify the upperbound in (5.33) as

$$\begin{aligned} \|x - x_{i+k}\|_A^2 &\le F_i^2 \sum_2^n \lambda_j \bar{P}_k(\lambda_j)^2 P_i(\lambda_j)^2 \mu_j^2 \\ &= F_i^2 \|x - \bar{x}_k\|_A^2 \\ &\le F_i^2 \frac{\|x - \bar{x}_k\|_A^2}{\|x - \bar{x}_0\|_A^2} \|x - x_i\|_A^2. \end{aligned} \tag{5.35}$$

We have now proved the result stated in Theorem 3.1 in [191]. The interpretation of this result is as follows. First note that when a Ritz value, say θ_1^i, is close to an eigenvalue, in this case λ_1, for some i then $F_i \approx 1$. The result in (5.35) says that we may then expect a reduction in the next steps that is bounded by the reduction that we obtain for a CG process for $Ax = b$ in which λ_1 is missing. If we define the effective condition number for that process as $\kappa_2 = \lambda_n/\lambda_2$, we have that

$$\|x - x_{i+k}\|_A \le 2 \left(\frac{\sqrt{\kappa_2} - 1}{\sqrt{\kappa_2} + 1} \right)^k \|x - x_i\|_A.$$

This shows that after the i-th step we have a reduction that is bounded by an expression in which we see the condition number for the remaining part of the spectrum. In fact, expression (5.35) shows more. It shows that we do not need to have accurate approximations for λ_1 by the first Ritz values. For instance, if

$$\frac{\theta_1^{(i)} - \lambda_1}{\lambda_1} < 0.1 \text{ and } \frac{\theta_1^{(i)} - \lambda_1}{\lambda_2 - \lambda_1} < 0.1,$$

then we already have that $F_i < 1.25$, which shows that from then on we have the same reduction as that for the reduced process except for a very modest factor. Things cannot then become worse, because it is easy to prove that $\theta_1^{(i)}$ is a strictly decreasing function towards λ_1 for increasing i. This means that F_i also becomes smaller during the further iterations.

Example We now return to our earlier example of Figure 5.3. For that example, we have at iteration step 10 a value $F_{10} \approx 1.22$. In Figure 5.4 we have plotted the errors in A-norm for the CG process (the drawn line), and we have also marked with $+$ the results for the comparison process. In this

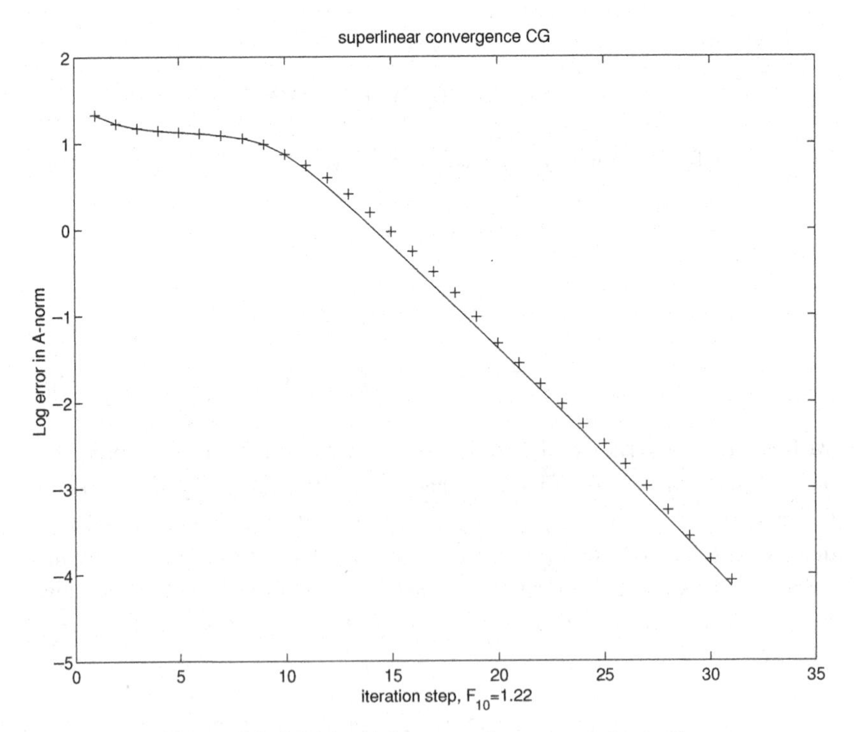

Figure 5.4. CG (–) and the comparison process (+ + +).

comparison process, we have removed at iteration step 10 the error component in the eigenvector direction corresponding to $\lambda_1 = 0.005$, and from then on we have multiplied the error-norms for the modified process by $F_1 0$. We see that this comparison process, which has only eigenvalues in the interval [0.08, 1.21], describes the observed convergence behaviour for the original CG process (with all eigenvalues present) rather well.

In [191] other results are presented that shed more light on all sorts of special situations, such as almost double eigenvalues in the spectrum of A, or situations in which more Ritz values have converged to a number of successive eigenvalues. Not surprisingly, these results show that after (modest) approximations of Ritz values for the smallest m eigenvalues, we will see convergence of the CG process from then on, governed by a condition number $\kappa_m \equiv \lambda_n / \lambda_{m+1}$. Of course, similar results can be derived for convergence of Ritz values towards the largest eigenvalues, but usually removal of small eigenvalues leads to much smaller condition numbers.

Exercise 5.23. *Consider the matrix A of order* 100 *with eigenvalues* 1, 2, ..., 100. *Suppose we carry out a CG process with this A, for some right-hand side b. What is the guaranteed speed of convergence after $\theta_1^{(i)} < 1.1$, for some i? How many eigenvalues at the upper end of the spectrum need to be well approximated by Ritz values in order to have a similar reduction with the effective condition number?*

In view of the fact that the convergence of Ritz values depends on the relative gap in the spectrum, we will see in the generic case that the smallest and the largest Ritz values converge about as fast as the smallest and largest eigenvalues of A, respectively. Is it meaningful in this to consider convergence of the largest Ritz value in order to explain an observed faster convergence?

5.4 CG and the normal equations

When faced with an unsymmetric system, it seems rather obvious to form the system of normal equations $A^T Ax = A^T b$, and to solve this iteratively with conjugate gradients. However, this may have severe disadvantages because of the squaring of the condition number. The effects of this are that the solution is more susceptible to errors in the right-hand side and that the rate of convergence of the CG procedure is much slower as it is for a comparable symmetric system with a matrix with the same condition number as A. Moreover, the amount of work per iteration step, necessary for the matrix vector product, is doubled. GMRES avoids our having to form the normal equations, but needs an increasing amount of storage and computational work per iteration. Therefore, the normal equations approach may still be attractive if only a few iterations suffice to obtain an acceptably small residual. To illustrate this, we note that when A is orthonormal, CG solves $A^T Ax = A^T b$ in one single iteration because $A^T A = I$. GMRES would need many iterations for such an extremely well-conditioned system if A is indefinite (which is the case for nontrivial situations). Of course, solving orthonormal systems is an easy matter, but this example shows that if A is in some sense close to orthogonal then the normal equations approach with CG may be very attractive. Another way of stating this is that CG with the normal equations, in the way formulated below in CGLS, may be attractive if the *singular values* of A are clustered. The Krylov subspace methods for the equation $Ax = b$ are attractive if the *eigenvalues of* A are clustered.

Paige and Saunders [154] point out that applying CG (see Figure 5.1, with A replaced by $A^T A$) naively to the normal equations is not advisable, largely because the method would then generate vectors of the form $A^T Ap_i$. This vector does not lead to sufficiently accurate iteration coefficients α_i. An algorithm with better numerical properties is obtained after a slight algebraic rearrangement, where we make use of the intermediate vector Ap_i (see also [28], which contains an early FORTRAN implementation). This leads to the CGLS algorithm in Figure 5.5. Our formulation is slightly different from the one in [154], in order

x_0 is an initial guess, $r_0 = b - Ax_0$, $s_0 = A^T r_0$
for $i = 1, 2, \ldots.$
  $\rho_{i-1} = s_{i-1}^T s_{i-1}$
  if $i = 1$
    $p_i = s_{i-1}$
  else
    $\beta_{i-1} = \rho_{i-1}/\rho_{i-2}$
    $p_i = s_{i-1} + \beta_{i-1} p_{i-1}$
  endif
  $q_i = Ap_i$
  $\alpha_i = \rho_{i-1}/q_i^T q_i$
  $x_i = x_{i-1} + \alpha_i p_i$
  $r_i = r_{i-1} - \alpha_i q_i$
  $s_i = A^T r_i$
  if x_i accurate enough **then** quit
end

Figure 5.5. CGLS.

to have it compatible with the CG algorithm in Figure 5.1. The vector r_i denotes the residual for the system $Ax = b$.

The use of conjugate gradients (LSQR) in a least squares context, as well as a theoretical comparison with SIRT type methods, is discussed in [192] and [193].

Exercise 5.24. *Show that, in exact arithmetic, CGLS leads to x_i for which $\|b - Ax_i\|_2$ is minimal over the shifted Krylov subspace $x_0 + \mathcal{K}^i(A^T A; A^T b)$.*

For overdetermined, or underdetermined, systems, solving the normal equations may be a quite natural approach. Now we have to face the situation that A may be ill-conditioned. In that case it may be numerically unattractive to apply the Lanczos procedure with $A^T A$ to form the reduced tridiagonal system with respect to the Krylov subspace. This tridiagonal system may also be ill-conditioned. Paige and Saunders have proposed a method that behaves numerically better. Their LSQR [154] is, in exact arithmetic, equivalent to CGLS but gives better results for ill-conditioned systems (in particular, when the singular vectors associated with small singular vectors carry important information about the solution). We will now discuss LSQR in more detail.

We first note that the least squares problem $Ax = b$ is equivalent to the auxiliary square linear system

$$\begin{pmatrix} I & A \\ A^T & 0 \end{pmatrix} \begin{pmatrix} r \\ x \end{pmatrix} = \begin{pmatrix} b \\ 0 \end{pmatrix}. \tag{5.36}$$

Forming an orthonormal basis for the Krylov subspace for the system in (5.36), and starting with the vector

$$w_1 = \frac{1}{\|b\|_2} \begin{pmatrix} b \\ 0 \end{pmatrix},$$

we obtain as the second vector for the Krylov subspace:

$$\frac{1}{\|b\|_2} \begin{pmatrix} b \\ A^T b \end{pmatrix}.$$

After orthogonalizing this vector to w_1 and normalizing the result, we obtain the second orthogonal basis vector

$$w_2 = \frac{1}{\|A^T b\|_2} \begin{pmatrix} 0 \\ A^T b \end{pmatrix}.$$

Exercise 5.25. *Show that we get alternately orthogonal basis vectors for the Krylov subspace of the form*

$$\begin{pmatrix} u \\ 0 \end{pmatrix} \quad \text{and} \quad \begin{pmatrix} 0 \\ v \end{pmatrix}. \tag{5.37}$$

Let U_k denote the matrix of the first k u-vectors and let V_k denote the matrix of the first k v-vectors in their proper order. Then we have the following relation between these vectors:

$$\beta_1 u_1 = b \quad \alpha_1 v_1 = A^T u_1$$

$$i = 1, 2, \ldots \quad \begin{cases} \beta_{i+1} u_{i+1} = A v_i - \alpha_i u_i \\ \alpha_{i+1} v_{i+1} = A^T u_{i+1} - \beta_{i+1} v_i \end{cases} \tag{5.38}$$

The scalars $\alpha_i > 0$ and $\beta_i > 0$ are chosen so that $\|u_i\|_2 = \|v_i\|_2 = 1$. These relations are also known as the bidiagonalization procedure, due to Golub and Kahan [95], for reasons that will now be made clear. With

$$\begin{array}{l} U_k \equiv [u_1, u_2, \ldots, u_k], \\ V_k \equiv [v_1, v_2, \ldots, v_k], \end{array} \quad B_k = \begin{bmatrix} \alpha_1 & & & \\ \beta_2 & \alpha_2 & & \\ & \beta_3 & \ddots & \\ & & \ddots & \alpha_k \\ & & & \beta_{k+1} \end{bmatrix},$$

it follows that

$$\beta_1 U_{k+1} e_1 = b, \tag{5.39}$$

$$A V_k = U_{k+1} B_k, \tag{5.40}$$

$$A^T U_{k+1} = V_k {B_k}^T + \alpha_{k+1} v_{k+1} e_{k+1}^T. \tag{5.41}$$

Exercise 5.26. *Show that $T_k = {B_k}^T B_k$, with T_k the reduced tridiagonal matrix obtained with the Lanczos process for $A^T A x = A^T b$, and that this Lanczos process is characterized by*

$$A^T A V_k = V_k T_k + \gamma_{k+1} v_{k+1} e_{k+1}^T, \tag{5.42}$$

with $\gamma = \beta_{k+1} \alpha_{k+1}$.

Now we return to the augmented system (5.36) (which is equivalent to the normal equations). With the Lanczos method, we want to find a solution in a

Krylov subspace $\mathcal{K}$ so that the residual is orthogonal to $\mathcal{K}$, that is, we want to find a $(\widetilde{r},\ \widetilde{x})^T \in \mathcal{K}$ so that

$$\begin{bmatrix} I & A \\ A^T & 0 \end{bmatrix} \begin{bmatrix} \widetilde{r} \\ \widetilde{x} \end{bmatrix} - \begin{bmatrix} b \\ 0 \end{bmatrix} \perp \mathcal{K}. \tag{5.43}$$

When we increase the dimension of $\mathcal{K}$ then $\widetilde{x}$ should approximate x and $\widetilde{r}$ should approximate $r = b - Ax$. In view of relation (5.42) the subspace $\mathcal{K}$ should be such that it permits $\widetilde{x}$ of the form $x_k \equiv \widetilde{x} = V_k y_k$. For the corresponding $r_k = b - Ax_k$, we should then have

$$\begin{aligned} r_k &= b - AV_k x_k \\ &= b - U_{k+1} B_k y_k \\ &= U_{k+1}(\beta_1 - B_k y_k) \\ &\equiv U_{k+1} t_{k+1}. \end{aligned}$$

This shows that $\mathcal{K}$ should be spanned by the columns of

$$\begin{bmatrix} U_{k+1} & 0 \\ 0 & V_k \end{bmatrix},$$

and note that these columns are generated by the first $2k + 1$ steps of the Lanczos process for the augmented system (5.36), although not in that order.

With $\widetilde{r} = U_{k+1} t_{k+1}$ and $\widetilde{x} = V_k y_k$, the orthogonality relations (5.43) lead to

$$\begin{bmatrix} U_{k+1}^T & 0 \\ 0 & V_k^T \end{bmatrix} \left[\begin{bmatrix} I & A \\ A^T & 0 \end{bmatrix} \begin{bmatrix} U_{k+1} t_{k+1} \\ V_k x_k \end{bmatrix} - \begin{bmatrix} b \\ 0 \end{bmatrix} \right] = 0.$$

This leads to the reduced system

$$\begin{bmatrix} I & B_k \\ B_k^T & 0 \end{bmatrix} \begin{bmatrix} t_{k+1} \\ y_k \end{bmatrix} = \begin{bmatrix} \beta_1 e_1 \\ 0 \end{bmatrix}. \tag{5.44}$$

Now note that (5.44) is equivalent to the least-squares problem

$$\min \|\beta_1 e_1 - B_k y_k\|_2, \tag{5.45}$$

and this forms the basis for LSQR.

In LSQR [154] the linear least-squares problem (5.45) is solved via the standard QR factorization (see, for instance, [98, Chapter 5]) for the matrix B_k.

$$
\begin{array}{l}
x_0 = 0,\ \beta_1 = \|b\|_2,\ u_1 = \|b\|_2/\beta_1, \\
v = A^T u_1, \alpha_1 = \|v\|_2,\ w_1 = v_1 = v/\alpha_1 \\
\widetilde{\phi}_1 = \beta_1, \widetilde{\rho}_1 = \alpha_1 \\
\textbf{for } i = 1, 2, \ldots. \\
\quad u = Av_i - \alpha_i u_i, \beta_{i+1} = \|u\|_2,\ u_{i+1} = u/\beta_{i+1} \\
\quad v = A^T u_{i+1} - \beta_{i+1} v_i, \alpha_{i+1} = \|v\|_2,\ v_{i+1} = v/\alpha_{i+1} \\
\quad \rho_i = \sqrt{\widetilde{\rho_i}^2 + \beta_{i+1}^2} \\
\quad c_i = \widetilde{\rho}_i/\rho_i \\
\quad s_i = \beta_{i+1}/\rho_i \\
\quad \theta_{i+1} = s_i \alpha_{i+1} \\
\quad \widetilde{\rho}_{i+1} = -c_i \alpha_{i+1} \\
\quad \phi_i = c_i \widetilde{\phi}_i \\
\quad \widetilde{\phi}_{i+1} = s_i \widetilde{\phi}_i \\
\quad x_i = x_{i-1} + (\phi_i/\rho_i) w_i \\
\quad w_{i+1} = v_{i+1} - (\theta_{i+1}/\rho_i) w_i \\
\quad \textbf{if } x_i \text{ accurate enough } \textbf{then} \text{ quit} \\
\textbf{end}
\end{array}
$$

Figure 5.6. LSQR.

The better numerical properties of LSQR, with respect to CG for the normal equations, stem from the fact that we form the bidiagonal factors of the Lanczos tridiagonal matrix and that we use the QR method for solving the reduced least-squares problem instead of solving the reduced tridiagonal problem (thus avoiding additional problems if this reduced tridiagonal system is ill-conditioned).

For completeness I give the LSQR algorithm, as formulated in [154], in Figure 5.6. For further details refer to [154] and note that more sophisticated implementations of LSQR are available in netlib:
`http://www.netlib.org`

An interesting variant that is also based on the bidiagonalization of A is the so-called Craig's method [154]. The easiest way to think of this method is to apply Conjugate Gradients to the system $A^T Ax = A^T b$, with the following choice for the inner product

$$[x, y] \equiv (x, (A^T A)^{-1} y),$$

which defines a proper inner product if A is of full rank (see Section 5.2).

First note that the two inner products in CG (as in Section 5.2) can be computed without inverting $A^T A$:

$$[p_i, A^T A p_i] = (p_i, p_i),$$

and, assuming that $b \in R(A)$ so that $Ax = b$ has a solution x:

$$\begin{aligned}[r_i, r_i] &= [A^T(Ax_i - b), A^T(Ax_i - b)] = [A^T A(x_i - x), A^T(Ax_i - b)] \\ &= (x_i - x, A^T(Ax_i - b)) = (Ax_i - b, Ax_i - b).\end{aligned}$$

Apparently, with CG we are minimizing

$$[x_i - x, A^T A(x_i - x)] = (x_i - x, x_i - x) = \|x_i - x\|_2^2,$$

that is, in this approach the Euclidean norm of the error is minimized, over the same shifted subspace $x_0 + K^i(A^T A, A^T r_0)$, as it is in the standard approach. Note, however, that the rate of convergence of Craig's method is also determined by the condition number of $A^T A$, so that this method is also only attractive if there is a good preconditioner for $A^T A$.

It is also possible to interpret Craig's method as a solution method for the system $AA^T u = b$, with $x = A^T u$. This leads to the same algorithm as sketched in the approach discussed here. For a convenient introduction to the AA^T-approach see [168].

5.5 Further references

A more formal presentation of CG, as well as many theoretical properties, can be found in the textbook by Hackbusch [111]. A shorter presentation is given in [98]. An overview of papers published in the first 25 years of existence of the method is given in [96]. Vector processing and parallel computing aspects are discussed in [61] and [148].

6

GMRES and MINRES

In this chapter I am still concerned with the problem of identifying good approximations x_i for the solution of $Ax = b$ in the Krylov subspace $\mathcal{K}^i(A, r_0)$, with $r_0 = b$ (I assume that $x_0 = 0$; the situation $x_0 \neq 0$ leads to a trivial shift of the approximations, cf. (3.6)). The construction of an orthogonal set of basis vectors v_j for the subspace $\mathcal{K}^i(A, r_0)$ leads to the relation (3.18):

$$AV_i = V_{i+1} H_{i+1,i}.$$

I will exploit this relation for the construction of approximations with minimum norm residual over the Krylov subspace.

6.1 GMRES

As we have seen in Section 4.2, the minimal norm residual approach leads to a small minimum least squares problem that has to be solved:

$$H_{i+1,i}\, y = \|r_0\|_2\, e_1. \tag{6.1}$$

In GMRES [169] this is done efficiently with Givens rotations, that annihilate the subdiagonal elements in the upper Hessenberg matrix $H_{i+1,i}$. The resulting upper triangular matrix is denoted by $R_{i+1,i}$:

$$Q_{i+1,i+1} H_{i+1,i} = R_{i+1,i}$$

where $Q_{i+1,i+1}$ denotes the product of the successive Givens eliminations of the elements $h_{j+1,j}$, for $j = 1, \ldots, i$.

Exercise 6.1. *Consider the 2 by 2 matrix*

$$A = \begin{bmatrix} a_{1,1} & a_{1,2} \\ a_{2,1} & a_{2,2} \end{bmatrix}.$$

Construct the 2 by 2 matrix

$$Q = \begin{bmatrix} \cos(\theta) & \sin(\theta) \\ -\sin(\theta) & \cos(\theta) \end{bmatrix},$$

so that $B = QA$ has a zero element in its lower left corner: $b_{2,1} = 0$. The matrix Q is called a Givens transformation. What is the value of θ? Verify that $Q^T Q = I$.

Show that Givens rotations can be used to reduce $H_{i+1,i}$ to upper triangular form. What do the Givens transformations look like?

After the Givens transformations the least squares solution y minimizes

$$\begin{aligned} \| H_{i+1,i} y - \|r_0\|_2 e_1 \|_2 &= \| Q_{i+1,i+1}^T R_{i+1,i} y - \|r_0\|_2 e_1 \|_2 \\ &= \| R_{i+1,i} y - Q_{i+1,i+1} \|r_0\|_2 e_1 \|_2. \end{aligned} \tag{6.2}$$

The resulting least squares problem leads to the minimum norm solution

$$y = R_{i,i}^{-1} Q_{i,i+1} \|r_0\|_2 e_1.$$

The required approximation x_i is now computed as $x_i = V_i y$.

Exercise 6.2. *Show that $R_{i,i}$ cannot be singular, unless x_{i-1} is equal to the solution of $Ax = b$.*

Note that when A is Hermitian (but not necessarily positive definite), the upper Hessenberg matrix $H_{i+1,i}$ reduces to a tridiagonal system. This simplified structure can be exploited in order to avoid storage of all the basis vectors for the Krylov subspace, in a similar way to that pointed out for CG. The resulting method is known as MINRES [153], see Section 6.4.

In order to avoid excessive storage requirements and computational costs for the orthogonalization, GMRES is usually restarted after each m iteration steps. This algorithm is referred to as GMRES(m); the not-restarted version is often called 'full' GMRES. There is no simple rule to determine a suitable value for

$$
\begin{array}{ll}
& r = b - Ax_0, \text{ for a given initial guess } x_0 \\
& x = x_0 \\
& \textbf{for } j = 1, 2, \ldots. \\
& \quad \beta = ||r||_2,\ v_1 = r/\beta,\ \widehat{b} = \beta e_1 \\
& \quad \textbf{for } i = 1, 2, \ldots, m \\
& \quad\quad w = Av_i \\
& \quad\quad \textbf{for } k = 1, \ldots, i \\
& \quad\quad\quad h_{k,i} = v_k^T w,\ w = w - h_{k,i} v_k \\
& \quad\quad h_{i+1,i} = \|w\|_2,\ v_{i+1} = w/h_{i+1,i} \\
& \quad\quad r_{1,i} = h_{1,i} \\
& \quad\quad \textbf{for } k = 2, \ldots, i \\
& \quad\quad\quad \gamma = c_{k-1} r_{k-1,i} + s_{k-1} h_{k,i} \\
& \quad\quad\quad r_{k,i} = -s_{k-1} r_{k-1,i} + c_{k-1} h_{k,i} \\
& \quad\quad\quad r_{k-1,i} = \gamma \\
& \quad\quad \delta = \sqrt{r_{i,i}^2 + h_{i+1,i}^2},\ c_i = r_{i,i}/\delta,\ s_i = h_{i+1,i}/\delta \\
& \quad\quad r_{i,i} = c_i r_{i,i} + s_i h_{i+1,i} \\
& \quad\quad \widehat{b}_{i+1} = -s_i \widehat{b}_i,\ \widehat{b}_i = c_i \widehat{b}_i \\
& \quad\quad \rho = |\widehat{b}_{i+1}| \ \ (= \|b - Ax_{(j-1)m+i}\|_2) \\
& \quad\quad \textbf{if } \rho \text{ is small enough } \textbf{then} \\
& \quad\quad\quad (n_r = i, \text{ goto } \mathit{SOL}) \\
& \quad n_r = m,\ y_{n_r} = \widehat{b}_{n_r}/r_{n_r,n_r} \\
\mathit{SOL}: & \quad \textbf{for } k = n_r - 1, \ldots, 1 \\
& \quad\quad y_k = (\widehat{b}_k - \sum_{i=k+1}^{n_r} r_{k,i} y_i)/r_{k,k} \\
& \quad x = x + \sum_{i=1}^{n_r} y_i v_i, \ \textbf{if } \rho \text{ small enough quit} \\
& \quad r = b - Ax
\end{array}
$$

Figure 6.1. Unpreconditioned GMRES(m) with modified Gram–Schmidt.

m; the speed of convergence may vary drastically for nearby values of m. It may be the case that GMRES($m + 1$) is much more expensive than GMRES(m), even in terms of numbers of iterations.

We present in Figure 6.1 the modified Gram–Schmidt version of GMRES(m) for the solution of the linear system $Ax = b$. The application to preconditioned systems, for instance $K^{-1}Ax = K^{-1}b$, is straightforward.

For complex valued systems, the scheme is as in Figure 6.2. Note that the complex rotation is the only difference with respect to the real version.

$$
\begin{array}{l}
r = b - Ax_0, \text{ for a given initial guess } x_0 \\
x = x_0 \\
\textbf{for } j = 1, 2, \ldots. \\
\quad \beta = ||r||_2, v_1 = r/\beta, \widehat{b} = \beta e_1 \\
\quad \textbf{for } i = 1, 2, \ldots, m \\
\quad\quad w = Av_i \\
\quad\quad \textbf{for } k = 1, \ldots, i \\
\quad\quad\quad h_{k,i} = v_k^* w, \ w = w - h_{k,i} v_k \\
\quad\quad h_{i+1,i} = ||w||_2, \ v_{i+1} = w/h_{i+1,i} \\
\quad\quad r_{1,i} = h_{1,i} \\
\quad\quad \textbf{for } k = 2, \ldots, i \\
\quad\quad\quad \gamma = c_{k-1} r_{k-1,i} + \bar{s}_{k-1} h_{k,i} \\
\quad\quad\quad r_{k,i} = -s_{k-1} r_{k-1,i} + c_{k-1} h_{k,i} \\
\quad\quad\quad r_{k-1,i} = \gamma \\
\quad\quad \delta = \sqrt{|r_{i,i}|^2 + |h_{i+1,i}|^2} \\
\quad\quad \textbf{if } |r_{i,i}| < |h_{i+1,i}| \\
\quad\quad \textbf{then } \mu = r_{i,i}/h_{i+1,i}, \ \tau = \bar{\mu}/|\mu| \\
\quad\quad \textbf{else } \mu = h_{i+1,i}/r_{i,i}, \ \tau = \mu/|\mu| \\
\quad\quad c_i = |r_{i,i}|/\delta, \ s_i = |h_{i+1,i}|\tau/\delta \\
\quad\quad r_{i,i} = c_i r_{i,i} + \bar{s}_i h_{i+1,i} \\
\quad\quad \widehat{b}_{i+1} = -s_i \widehat{b}_i, \ \widehat{b}_i = c_i \widehat{b}_i \\
\quad\quad \rho = |\widehat{b}_{i+1}| \ \ (= ||b - Ax_{(j-1)m+i}||_2) \\
\quad\quad \textbf{if } \rho \text{ is small enough } \textbf{then} \\
\quad\quad\quad (n_r = i, \text{ goto } SOL) \\
\quad n_r = m, \ y_{n_r} = \widehat{b}_{n_r}/r_{n_r,n_r} \\
SOL: \quad \textbf{for } k = n_r - 1, \ldots, 1 \\
\quad\quad y_k = (\widehat{b}_k - \sum_{i=k+1}^{n_r} r_{k,i} y_i)/r_{k,k} \\
\quad x = x + \sum_{i=1}^{n_r} y_i v_i, \ \textbf{if } \rho \text{ small enough quit} \\
\quad r = b - Ax
\end{array}
$$

Figure 6.2. Unpreconditioned GMRES(m) for complex systems.

Exercise 6.3. *For a real unsymmetric nondefective matrix A there exists a nonsingular real matrix Z so that*

$$Z^{-1}AZ = T,$$

where T is a real block diagonal matrix with 2 by 2 blocks along the diagonal. The eigenvalues of A are equal to those of T.

This can be used to construct test examples for GMRES. Would it be sufficient to test GMRES with diagonal matrices, as was the case for Conjugate Gradients? Why not?

Exercise 6.4. *The FOM method for real unsymmetric systems is based on the Galerkin relations:*

$$b - Ax_m \perp \mathcal{K}^m(A, x_0),$$

for $x_m \in \mathcal{K}^m(A, x_0)$.

Modify the scheme for (the real version of) GMRES(m) so that we obtain the restarted version of FOM: FOM(m).

Exercise 6.5. *Test the methods GMRES(m) and FOM(m). Run the tests with only one cycle of m steps. Construct matrices with complex conjugate eigenvalues with positive real part.*

Plot the logarithm of the norms of the first m residuals. Why do the norms for GMRES form a strictly monotonic decreasing curve? Should that also be the case for FOM? Do you observe situations where $\|r_i^{GMRES}\|_2 \approx \|r_i^{FOM}\|_2$ *for some values of i? Could such situations be expected after inspection of the matrix* $H_{i+1,i}$*?*

Exercise 6.6. *Repeat the experiments for GMRES and FOM with real indefinite matrices (eigenvalues with positive real parts and eigenvalues with negative real parts). Can we modify a given test example so that FOM breaks down at step i? Hint: If* $H_{i+1,i}$ *is the reduced matrix for A, what is the reduced matrix for* $A - \sigma I$*? How can we expect a breakdown step from inspection of* $R_{i,i}$.

What happens to GMRES at a breakdown step of FOM? Can FOM be continued after a breakdown? Modify FOM so that it computes only the approximated solutions at nonbreakdown steps.

Another scheme for GMRES, based upon Householder orthogonalization instead of modified Gram–Schmidt has been proposed in [215]. For certain applications it seems attractive to invest in additional computational work in return for improved numerical properties: the better orthogonality might save iteration steps.

The eigenvalues of $H_{i,i}$ are the Ritz values of A with respect to the Krylov subspace spanned by $v_1, \ldots, v_i$. They approximate eigenvalues of A increasingly well for increasing dimension i.

Exercise 6.7. *Construct some example systems for the testing of GMRES, such that the eigenvalues of the matrices are known. Plot the convergence curve:* $\log(\|r_i\|_2)$ *as a function of* i. *Compute also the Ritz values for some values of* i. *Is there any relation between the Ritz values and the eigenvalues of the matrix of the linear system? Which eigenvalues are well approximated? Does that depend on the distribution of the eigenvalues of the matrix? Is there a relation between the convergence of Ritz values and effects in the convergence curve?*

There is an interesting and simple relation between the Ritz–Galerkin approach (FOM and CG) and the minimum residual approach (GMRES and MINRES). In GMRES the projected system matrix $H_{i+1,i}$ is transformed by Givens rotations to an upper triangular matrix (with last row equal to zero). So, in fact, the major difference between FOM and GMRES is that in FOM the last ($(i+1)$-th) row is simply discarded, while in GMRES this row is rotated to a zero vector. Let us characterize the Givens rotation, acting on rows i and $i+1$, in order to zero the element in position $(i+1, i)$, by the sine s_i and the cosine c_i. Let us further denote the residuals for FOM with a superscript F and those for GMRES with a superscript G. If $c_k \neq 0$ then the FOM and the GMRES residuals are related by

$$\|r_k^F\|_2 = \frac{\|r_k^G\|_2}{\sqrt{1 - (\|r_k^G\|_2 / \|r_{k-1}^G\|_2)^2}}, \tag{6.3}$$

([49, Theorem 3.1]).

From this relation we see that when GMRES has a significant reduction at step k, in the norm of the residual (i.e., s_k is small, and $c_k \approx 1$), FOM gives about the same result as GMRES. On the other hand when FOM has a breakdown ($c_k = 0$), GMRES does not lead to an improvement in the same iteration step. Because of these relations we can link the convergence behaviour of GMRES with the convergence of Ritz values (the eigenvalues of the 'FOM' part of the upper Hessenberg matrix). This has been exploited in [206], for the analysis and explanation of local effects in the convergence behaviour of GMRES. We will see more of this in Section 6.2.

There are various methods that are mathematically equivalent to FOM or GMRES. We will say that two methods are mathematically equivalent if they produce the same approximations $\{x_k\}$ in exact arithmetic. Among those that are equivalent to GMRES are: ORTHOMIN [213], Orthodir [121], GENCR [75], and Axelsson's method [9]. These methods are often more expensive than GMRES per iteration step and they may also be less robust.

The ORTHOMIN method is still popular, since this variant can be easily truncated (ORTHOMIN(s)), in contrast to GMRES. The truncated and restarted versions of these algorithms are not necessarily mathematically equivalent.

Methods that are mathematically equivalent to FOM are: Orthores [121] and GENCG [42, 221]. In these methods the approximate solutions are constructed such that they lead to orthogonal residuals (which form a basis for the Krylov subspace; analogously to the CG method).

The GMRES method and FOM are closely related to vector extrapolation methods, when the latter are applied to linearly generated vector sequences. For a discussion on this, as well as for implementations for these matrix free methods, see [172]. For an excellent overview of GMRES and related variants, such as FGMRES, see [168].

6.1.1 A residual vector variant of GMRES

It may not come as a surprise that a method as popular as GMRES comes in different flavours. As an example of such a variant we discuss the so-called *Simpler GMRES*, proposed by Walker and Zhou [216]. The main idea is that GMRES computes the vector from the Krylov subspace for which the residual had minimum norm. The residuals live in the shifted Krylov supspace

$$r_0 + A\mathcal{K}^m(A; r_0),$$

so that we may construct an orthonormal basis for $A\mathcal{K}^m(A; r_0)$ as an alternative to the regular basis.

More precisely, given a starting vector x_0 with residual vector r_0, the component z of the approximated solution $x_0 + z$ is sought in the subspace $\mathcal{B}$ spanned by $r_0, Ar_0, \ldots, A^{m-1}r_0$. The residual for z is computed with Az and this vector belongs to the subspace $\mathcal{V}$ spanned by the vectors $Ar_0, A^2r_0, \ldots, A^m r_0$. An orthonormal basis $v_1, v_2, \ldots, v_m$ is computed as in Figure 6.3.
Now note that the vectors $r_0, v_1, \ldots, v_{m-1}$ form a (nonorthogonal) basis for $\mathcal{B}$. We denote the matrix with columns $r_0, v_1, \ldots, v_{m-1}$ as B_m.

The vectors $v_1, \ldots, v_m$ form an orthonormal basis for $\mathcal{V}$, the matrix with these columns is denoted as V_m.

Exercise 6.8. *Show that*

$$AB_m = V_m U_m, \tag{6.4}$$

with U_m an m by m upper triangular matrix with elements $u_{i,j}$ (cf. Figure 6.3).

```
r_0 is a given vector, v_0 = r_0
for k = 1, ..., m
        w = A v_{k-1}
        for j = 1, ..., k - 1
                u_{j,k} = v_j^T w, w = w - u_{j,k} v_j
        end j
        u_{k,k} = ||w||_2, v_k = w / h_{k,k}
end k
```

Figure 6.3. A basis for $A\mathcal{K}^m(A; r_0)$ with modified Gram–Schmidt.

Now we are well equipped to compute the residual minimizing approximation $x_0 + z$ from the shifted subspace $x_0 + \mathcal{B}$. This z can be expressed as $z = B_m s$, which gives the approximated solution $x_0 + B_m s$. The vector s has to be determined so that the residual in 2-norm

$$\|b - A(x_0 + B_m s)\|_2$$

is minimal. Note that the residual for this approximation is $r_m = b - A(x_0 + B_m s)$. Using (6.4) we obtain

$$\begin{aligned} \|b - A(x_0 + B_m s)\|_2 &= \|r_0 - AB_m s\|_2 \\ &= \|r_0 - V_m U_m s\|_2. \end{aligned} \tag{6.5}$$

For the minimization of the last expression in (6.5), we have to decompose r_0 into a component in $\mathcal{V}$ and a vector orthogonal to $\mathcal{V}$:

$$r_0 = t + V_m w,$$

with $w = V_m^T r_0$. This leads to

$$\|r_0 - V_m U_m s\|_2 = \|t + V_m(w - U_m s\|_2, \tag{6.6}$$

and it follows that the minimal norm is achieved for the solution s of

$$U_m s = w. \tag{6.7}$$

Exercise 6.9. *Prove that $r_m = t$.*

With $s = U_m^{-1} w$ (6.7), the approximated solution from $\mathcal{B}$ is obtained as

$$x_m = x_0 + B_m s.$$

We are now ready to compare this variant of GMRES with the original GMRES, cf. Figure 6.1. In the traditional form we work with a basis for the standard Krylov subspace, onto which A is reduced to upper Hessenberg form. This leads to the solution of a small upper Hessenberg system, see (6.1). For the variant in this section, we have to solve a small upper triangular system. That means that we have avoided the Givens rotations, necessary for the reduction of the upper Hessenberg system to an upper triangular system. For this reason the new variant has been named *simpler* GMRES.

The question now is: what have we actually saved and what is the price to be paid? First we compare the costs for original GMRES and *simpler* GMRES. The only difference is in the amount of scalar work; the operations with (long) vectors and with A are the same. GMRES additionally requires 16 flops per iteration and one square root. *Simpler* GMRES requires 2 flops per iteration plus sin and an arccos (for the update of the norm of the residual). This means that both methods have about the same expense per iteration.

Simpler GMRES has available in each iteration the residual vector, and that means that we do not have to compute it explicitly at a restart (as in GMRES(m)). However, the residuals in Simpler GMRES are obtained by an update procedure (projecting out the new basis vector for $\mathcal{V}$) and hence the approximate solution and the residual vector, which should be equal in exact arithmetic, may deviate after a few iterations. So, unless the length of a GMRES cycle is very small, it may be better to compute the residual explicitly at restart, leading to only a small increase of the total computational costs.

It seems to be more serious that the approximate solution is computed for the nonorthogonal basis of $\mathcal{B}$. It was shown in [131] that

$$\kappa_2(B_m) \leq 2\frac{\|r_0\|_2 + (\|r_0\|_2^2 - \|r_m\|_2^2)^{1/2}}{\|r_m\|_2}.$$

In [216] an upperbound was given, with essentially the same order of magnitude. This implies that fast convergence goes hand in hand with an ill-conditioned basis, which may lead to inaccuracies in the computed approximation. A similar effect does not play a role for slowly converging restarted *simpler* GMRES processes [216], but there may be negative effects for larger values of m, as numerical experiments in [131] show.

In conclusion it seems that *simpler* GMRES is not a serious competitor for original GMRES, but it may be that the process has its merits in situations with low values of m, for instance in hybrid iteration methods (such as GMRESR or FGMRES).

6.1.2 Flexible GMRES

As we have seen in Section 6.1.1, it is necessary to have an orthonormal basis for the image subspace, that is the subspace in which the residuals 'live', in order to determine the minimum norm residual. Saad [166] has exploited this knowledge for the design of a variant of GMRES that admits variable preconditioning, more particularly right preconditioning. That is, we consider solving the linear system

$$AK^{-1}y = b, \tag{6.8}$$

with $x = K^{-1}b$. The application of GMRES with the operator AK^{-1} is straightforward. Note that the approximate solution for y has to be multiplied by K^{-1} in order to get a meaningful approximation for x. This operation can be saved by storing all the intermediate vectors $K^{-1}v_i$ that have to be computed when forming the new basis vector through $w = AK^{-1}v_i$. Thus one operation might be saved with the preconditioner per m steps at the expense of storing m long vectors. Often storage is a limiting factor and this will not then be attractive. However, as Saad notes [168, p. 257], there are situations where the preconditioner is not explicitly given, but implicitly defined via some computation, for instance as an approximate Jacobian in a Newton iteration, or by a few steps of an iteration process. Another example is when the preconditioning step is done by domain decomposition and if the local problems per domain are again solved by a few steps of an iterative method.

We denote this variable preconditioning by an index, that is in iteration step i we compute $w = AK_i^{-1}v_i$ and now we have to store the intermediate vectors $z_i = K_i^{-1}v_i$ in order to compose the update for the approximate solution x.

If we follow the Arnoldi process for the new vectors $w = AK_i^{-1}v_i$, then we do not obtain a Krylov subspace, because the operator AK_i^{-1} is not fixed.

Exercise 6.10. *Let the Arnoldi process be carried out with the operator AK_j^{-1} for the construction of the orthogonal vector $\widetilde{v}_{j+1}$ and start with the initial normalized vector $\widetilde{v}_1$. Let Z_m denote the matrix with columns $z_j = K_j^{-1}\widetilde{v}_j$, then show that*

$$AZ_m = \widetilde{V}_{m+1}\widetilde{H}_{m+1,m}, \tag{6.9}$$

with $\widetilde{H}_{m+1,m}$ an $m+1$ by m upper Hessenberg matrix.

The residual minimization process can now be carried out in the same way as that for GMRES, with $\widetilde{H}_{m+1,m}$ instead of $h_{m+1,m}$. Of course, the update for x is with the z_j vectors. The complete FGMRES(m) scheme is given in Figure 6.4.

$r = b - Ax_0$, for a given initial guess x_0
$x = x_0$
for $j = 1, 2, \ldots.$
  $\beta = ||r||_2$, $\widetilde{v}_1 = r/\beta$, $\widehat{b} = \beta e_1$
  for $i = 1, 2, \ldots, m$
    $z_i = K_i^{-1}\widetilde{v}_i$
    $w = Az_i$
    for $k = 1, \ldots, i$
      $\widetilde{h}_{k,i} = \widetilde{v}_k^T w$, $w = w - \widetilde{h}_{k,i}\widetilde{v}_k$
    $\widetilde{h}_{i+1,i} = ||w||_2$, $\widetilde{v}_{i+1} = w/\widetilde{h}_{i+1,i}$
    $r_{1,i} = \widetilde{h}_{1,i}$
    for $k = 2, \ldots, i$
      $\gamma = c_{k-1}r_{k-1,i} + s_{k-1}\widetilde{h}_{k,i}$
      $r_{k,i} = -s_{k-1}r_{k-1,i} + c_{k-1}\widetilde{h}_{k,i}$
      $r_{k-1,i} = \gamma$
    $\delta = \sqrt{r_{i,i}^2 + \widetilde{h}_{i+1,i}^2}$, $c_i = r_{i,i}/\delta$, $s_i = \widetilde{h}_{i+1,i}/\delta$
    $r_{i,i} = c_i r_{i,i} + s_i\widetilde{h}_{i+1,i}$
    $\widehat{b}_{i+1} = -s_i\,\widehat{b}_i\ \widehat{b}_i = c_i\,\widehat{b}_i$
    $\rho = |\widehat{b}_{i+1}|\ \ (= ||b - Ax_{(j-1)m+i}||_2)$
    if ρ is small enough **then**
      $(n_r = i$, goto *SOL*)
  $n_r = m$, $y_{n_r} = \widehat{b}_{n_r}/r_{n_r,n_r}$
SOL: **for** $k = n_r - 1, \ldots, 1$
    $y_k = (\widehat{b}_k - \sum_{i=k+1}^{n_r} r_{k,i}y_i)/r_{k,k}$
  $x = x + \sum_{i=1}^{n_r} y_i z_i$, **if** ρ small enough quit
  $r = b - Ax$

Figure 6.4. FGMRES(m) with variable right preconditioning K_i^{-1}.

Relation (6.9) reveals the possible problems that we may encounter with the flexible approach. The matrix Z_m is nonorthogonal and the matrix $\widetilde{H}_{m+1,m}$ is not even an orthogonal projection of some right-preconditioned A (unless all preconditioners K_i are equal). That means that the reduced matrix $H_{m+1,m}$ may be singular even if A is nonsingular. This situation is rare, but it should be checked. From [168, Property 9.3] we have that the iterative process has arrived at the exact solution $x_j = x$ when the orthogonal basis v_j cannot be further expanded (i.e. when $h_{j+1,j} = 0$), under the condition that $H_{j,j}$ is nonsingular.

In Section 6.6 we will see another variant of GMRES that permits variable preconditioning. In that variant the updates for the residuals are preconditioned rather than the updates for the approximate solution as in FGMRES.

6.2 The convergence behaviour of GMRES

For CG we have seen nice upperbounds for the reduction of the residual norms $\|r_k\|_2/\|r_0\|_2$ in terms of the condition number of the matrix A (cf. (5.25). The situation for GMRES is much more complicated, because of the fact that unsymmetric matrices have, in general, no orthonormal eigensystem and the eigenvalues may be complex. From the Krylov subspace characterization it follows that the residual r_k, for full unpreconditioned GMRES, can be expressed in polynomial form as

$$r_k = P_k(A)r_0,$$

with $P_k(0) = 1$. Now assume that A is diagonalizable, that is there exists a nonsingular X such that

$$A = XDX^{-1},$$

with D a diagonal matrix with the (possibly complex) eigenvalues of A on its diagonals. We assume that the columns of X are scaled to make $\|X\|_2\|X^{-1}\|_2$ as small as possible. GMRES generates implicitly the polynomial P_k for which $\|r_k\|_2$ is minimal. Hence

$$\begin{aligned}\|r_k\|_2 &= \min_{P_k, P_k(0)=1} \|P_k(A)r_0\|_2 \\ &\leq \|X\|_2\|X^{-1}\|_2 \min_{P_k, P_k(0)=1} \max_j |P_k(\lambda_j)|\|r_0\|_2. \end{aligned} \qquad (6.10)$$

The upperbound given by (6.10) is, in general, not very useful. The bound is not sharp [101, Section 3.2], and, moreover, the minimization of a polynomial over a complex set of numbers, under the condition that it is 1 at the origin, is an unsolved problem. In fact, the upperbound does not even predict that $\|r_k\|_2 \leq \|r_0\|_2$, as is the case for GMRES. In some cases, the upperbound may be used to predict a true reduction for the GMRES residuals. For instance, when the *field of values* $\mathcal{F}(A)$ (Definition 2.2) of A is contained in an ellipse in the right-half plane then the rate of convergence can be bounded. Let us assume that A is real, so that the eigenvalues appear in complex conjugate pairs. The ellipse then has its centre d at the real axis, say $d > 0$. We denote the focal points by $d - c$ and $d + c$ and the intersection of the ellipse with the real axis by $d - a$ and $d + a$. Manteuffel [135] describes a Chebyshev iteration for $Ax = b$, and gives, for the situation described above, an asymptotic convergence factor

$$r_c = \frac{a + \sqrt{a^2 - c^2}}{d + \sqrt{d^2 - c^2}}.$$

Of course, there are infinitely many ellipses that contain $\mathcal{F}(A)$, but we have to select the smallest one. In [135] an algorithm is presented that computes the best parameters a, c, and d.

Because the GMRES residuals are smaller than the Chebyshev residuals, the GMRES residuals can be bounded by a set of numbers that decrease geometrically with this r_c. Note that because of the appearance of the condition number of X this does not predict a true reduction if this condition number is larger than $(1/r_c)^k$.

In general, nothing can be said about the convergence behaviour of GMRES, and no reduction can even be guaranteed for $k < n$, as the example in the next exercise shows.

Exercise 6.11. *Let e_i denote the canonical basis vectors in $\mathbb{R}^n$. Let A be the matrix with successive columns $e_2, e_3, \ldots, e_n, e_1$. For b we take $b = e_1$. We start full GMRES with $x_0 = 0$. Show that the upper Hessenberg matrices associated with the Arnoldi basis for the Krylov subspaces of dimension less than n have upper triangular part equal to zero. Use this in order to show that $\|r_j\| = \|r_0\|_2$ for all $j \leq n$. What happens at the n-th iteration step? Where are the eigenvalues of A located, and what are the Ritz values (the eigenvalues of the Hessenberg matrices)?*

The remarkable observation is that the condition number of the matrix in Exercise 6.11 is 1. This also shows that the convergence of GMRES cannot adequately be described in terms of the condition number. It also cannot, in general, be described in terms of the eigenvalues, as has been shown in [102]. The main result for convergence is given in the next theorem.

Theorem 6.1. *Given a nonincreasing positive sequence $f_0 \geq f_1 \geq f_{n-1}$ and a set of nonzero complex numbers $\lambda_1, \lambda_2, \ldots, \lambda_n$, there exists a matrix A with eigenvalues λ_j and a right-hand side b with $\|b\|_2 = f_0$ such that the residual vectors r_k of GMRES (for $Ax = b$, with $x_0 = 0$) satisfy $\|r_k\|_2 = f_k$, $k = 0, 1, \ldots, n-1$.*

So, the eigenvalue information alone is not enough; information about the eigensystem is also needed. If the eigensystem is orthogonal, as for *normal* matrices, then the eigenvalues are descriptive for the convergence. Also for well-conditioned eigensystems the distribution of the eigenvalues may give some insight into the actual convergence behaviour of GMRES. In practical cases *superlinear convergence behaviour* is also observed, as it is for CG. This has been explained in [206]. In that paper it is shown that when eigenvalues of the Hessenberg matrix $H_{k,k}$ (the Ritz values) approximate eigenvalues of A rather well, then the convergence of GMRES proceeds from then as if the corresponding eigenvector of A is no longer present. The analysis in [206] is unnecessarily complicated, because the GMRES approximations are computed with $H_{k+1,k}$. The approximations of FOM, however, are determined by $H_{k,k}$ and, hence, the FOM iteration polynomial can be expressed in terms of the Ritz polynomial (cf. (5.31)). Since the convergence of GMRES can be linked with that of FOM, the link with the Ritz values can be made. The introduction of the FOM polynomial could have been avoided by replacing the Ritz values by the *Harmonic Ritz values* [152], because the GMRES iteration polynomial can be characterized by the Harmonic Ritz values ([86]).

Embree [77] has studied the convergence behaviour of GMRES and GMRES(m) in great detail. He has studied upperbounds in terms of eigenvalues and in terms of fields of values, similar to the bounds we have seen here, and in terms of pseudospectral information. The latter bounds seem to be rather promising and a nice aspect is that relevant pseudospectral information can be gathered from the GMRES iteration process. This is certainly an interesting direction for further research.

6.3 Some numerical illustrations

My numerical experiments have been taken from [202]. The matrices A are of order = 200 and of the form $A = SBS^{-1}$ with

$$S = \begin{bmatrix} 1 & \beta & & & \\ & 1 & \beta & & \\ & & \ddots & \ddots & \\ & & & \ddots & \beta \\ & & & & 1 \end{bmatrix} \quad \text{and} \quad B = \begin{bmatrix} \lambda_1 & & & & \\ & \lambda_2 & & & \\ & & \lambda_3 & & \\ & & & \ddots & \\ & & & & \lambda_n \end{bmatrix}.$$

We take b such that the solution x of $Ax = b$ is S times the vector with all elements equal to 1. The value of β can be used to make A less normal. Note

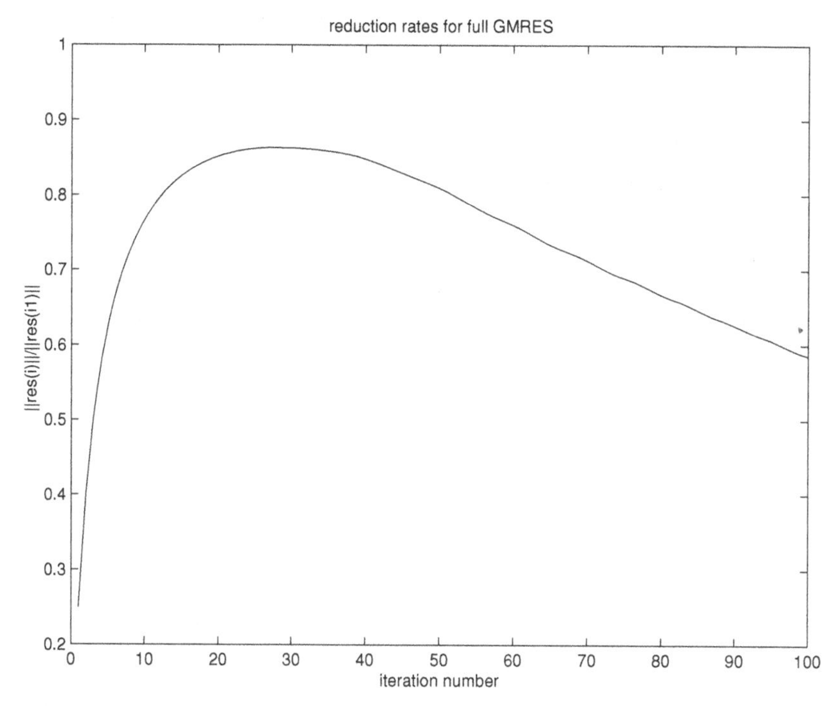

Figure 6.5. Successive reduction factors for the residual norms of GMRES for a uniform real spectrum.

that for $\beta = 0$, we have a very simple normal and even symmetric situation and that is not of so much interest for GMRES.

Example 1. For the first example we select $\beta = 0.9$ and $\lambda_j = j$, for $j = 1, 2, \ldots, n$. In Figure 6.5 we have shown the reduction factors $\|r_j\|_2/\|r_{j-1}\|_2$, for $j = 1, \ldots$.

We see that initially there is rather fast convergence, but this slows down until GMRES 'sees' the full spectrum of A. Then we see a maximum at about iteration numbers 26–27. At iteration 25, the smallest Ritz value $\theta_1^{(25)} \approx 2.15$, and at iteration 27 it is $\theta_1^{(27)} \approx 1.93$. In line with [206], the GMRES process can, from iteration 27 onwards, be compared with the residual norms of r_i' of a GMRES process, in which the first eigenvalue (and eigenvector) are missing:

$$\|r_{27+i}\|_2 \leq \kappa_2(S)\frac{|\theta_1^{(i+27)}|}{|\lambda_1|} \max_{j\geq 2} |\frac{\lambda_j - \lambda_1}{\lambda_j - \theta_1^{27+i}}| \|r_i'\|_2 \tag{6.11}$$

(cf. (5.35) for CG).

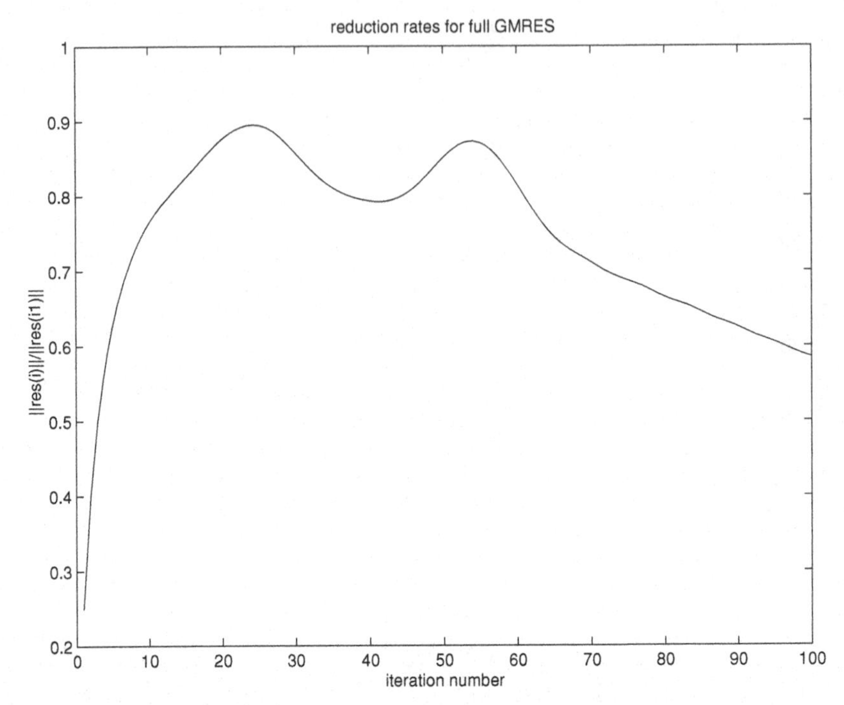

Figure 6.6. Successive reduction factors for the residual norms of GMRES for a close eigenpair.

After iteration 27 we observe a reduction in the rate of convergence, and this reduction continues to decrease, because, for this problem, next Ritz values $\theta_k^{(j)}$ also arrive in their intervals $[\lambda_k, \lambda_{k+1}]$. This explains the increasingly fast convergence.

Example 2. For the second example, λ_j is replaced by the value 1.1, which leads to a matrix with two relatively close eigenvalues.

In Figure 6.6 we see the ratios of the successive residual norms for this example. We see a first bulge in the graph of the reduction factors. After this bulge we see a phase of decreasing reduction factors, which has its minimum at about iteration number 40. At that point the smallest Ritz value $\theta_1^{(40)} \approx 1.0674$, which indicates that the underlying Arnoldi process has spotted the close eigenpair and has located a Ritz value near the middle. Then the process has to identify both eigenvalues and this explains the second bulge: at iteration 60 we find the two smallest Ritz values $\theta_1^{(60)} \approx 1.0326$ and $\theta_2^{(60)} \approx 1.2424$. This means that both Ritz values have arrived in their final intervals and from then on we may expect faster convergence, just as for the previous example.

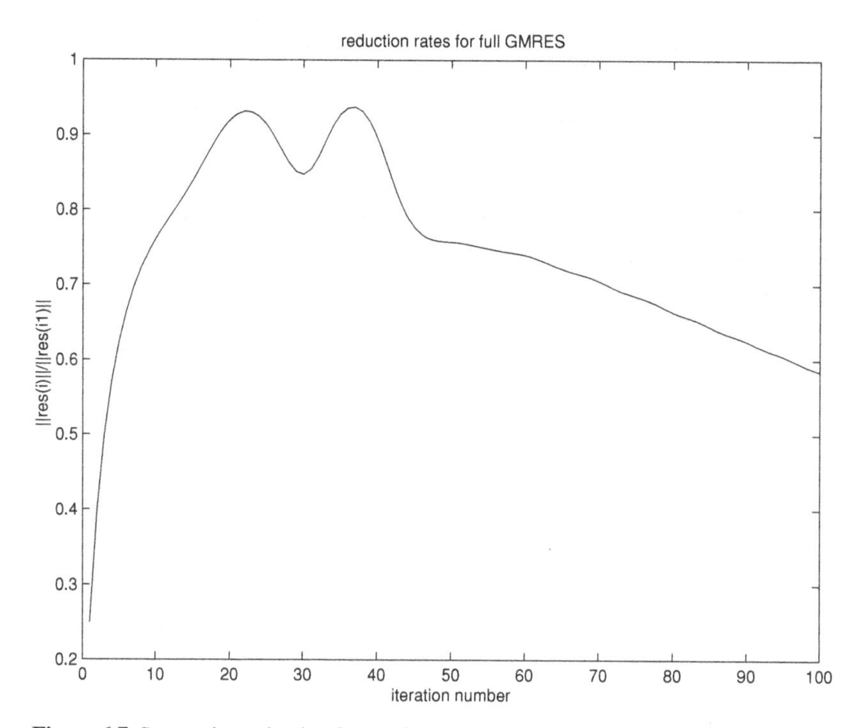

Figure 6.7. Successive reduction factors for the residual norms of GMRES for a complex conjugate eigenpair.

Example 3. We start again with the B of Example 1, but now the diagonal subblock

$$\begin{matrix} 1 & 0 \\ 0 & 2 \end{matrix}$$

is replaced by the subblock

$$\begin{matrix} 1 & 1 \\ -1 & 1 \end{matrix},$$

that is, the eigenvalues 1 and 2 of Example 1 are replaced by the conjugate eigenpair $1 \pm i$. In Figure 6.7 we see the reduction factors for successive residual norms for this example. We see a behaviour similar to that for the close eigenpair situation in Example 2. Indeed, at iteration 30 the smallest Ritz value is $\theta_1^{(30)} \approx 1.0580$ (the second one is $\theta_2^{(40)} \approx 4.0783$). The first one has arrived close to the middle of the conjugate eigenpair, while the second one is still not in its final interval $[3, 4]$. At iteration 40 we have $\theta_1^{(40)} \approx 0.95 + i$ and $\theta_2^{(40)} \approx 0.95 - i$

(in this case the next nearest Ritz value is $\theta_3^{(40)} \approx 3.82$. Apparently, the process had discovered the conjugate eigenpair and it has even a Ritz value in the next interval [3, 4]. With repeated applications of formulas like (6.11) we can explain the faster convergence after the second bulge.

Example 4. The final example is one with a defective matrix. We replace the leading 3 by 3 block of B (of Example 1) by a Jordan block with multiple eigenvalue 1:

$$\begin{matrix} 1 & 1 & 0 \\ 0 & 1 & 1. \\ 0 & 0 & 1 \end{matrix}$$

For β we again take the value 0.9. It should be noted that other values of β give more or less similar results, although the condition number of S increases for larger β. This is reflected by slower convergence of GMRES. For values of β significantly larger than 1.0 the condition number of S becomes very large and then there is practically no more convergence. In view of the large condition number of A the approximated solution would then also bear little significance and solution of such systems would be meaningless.

In Figure 6.8 we see the reduction factors for successive residual norms. At the first bulge we have the situation that a Ritz value approaches 1.0, at the second bulge a second Ritz value comes close to 1.0 and it is only after a third Ritz value has arrived near 1.0 (the third bulge) that faster convergence can take place. This faster convergence is in line with what may be expected for a system from which the Jordan block has been deleted. We can make two remarks for this example. First, it can be argued that defective matrices do not occur in rounding arithmetic. This may be so, but if the transformation matrix X is computed for transformation to diagonal form then X has a very large condition number, which also leads to slow convergence. Second, the situation in this example is essentially different from the situation of a diagonalizable matrix with multiple eigenvalue 1. As we have seen, the multiplicity of an eigenvalue plays no role in the nondefective case, since all eigenvector components corresponding to the multiple eigenvalue can be eliminated with one single root of the iteration polynomial P_k.

Comments For a more precise discussion on similar examples see [206, Section 3].

Although our experiments have been carried out for very simple examples, we see phases of convergence behaviour that are also typical for real

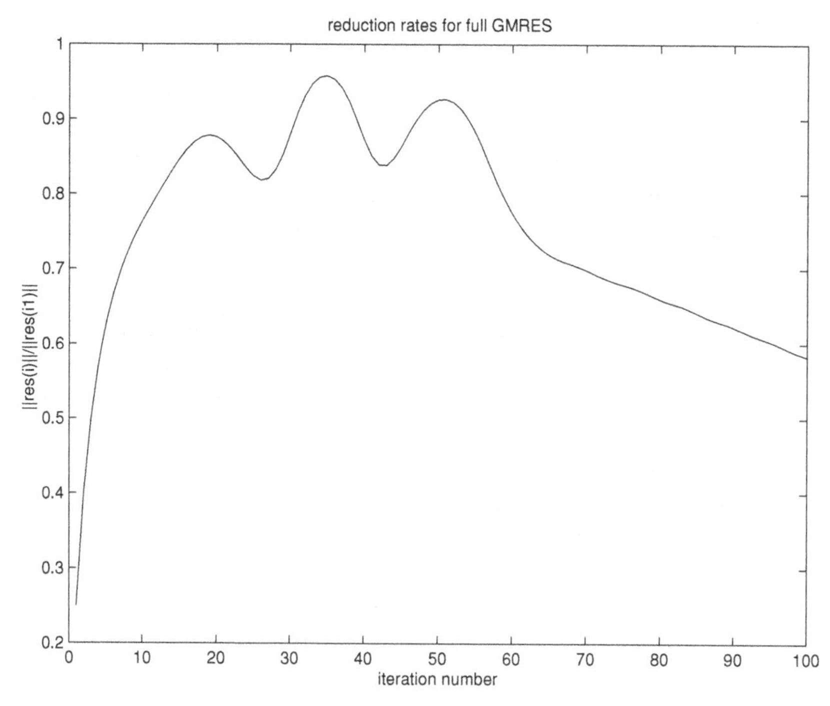

Figure 6.8. Successive reduction factors for the residual norms of GMRES for a defective matrix.

life problems. This type of analysis helps us to understand when and why preconditioning may be effective. With preconditioning matrix K we hope to cluster the eigenvalues of $K^{-1}A$ and to have well-separated eigenvalues remaining. In the situation that a large number of similar problems have to be solved, then my advice is to select one representative system from this group and start with GMRES, rather than one of the more memory friendly methods like Bi-CGSTAB. Then do as many steps with full GMRES as memory space and time permit and inspect the Hessenberg system computed in GMRES. Compute the Ritz values of some successive Hessenberg matrices $H_{k,k}$, which most often gives a good impression of the spectrum of the (preconditioned) matrix A. This gives an indication of the possible effectiveness of preconditioning and if the spectrum is nicely located, more memory friendly Krylov methods may be considered. If the spectrum leads us to expect slow convergence, for instance, because it has the origin in its convex hull, then my advice is to seek a better preconditioner (or to expect the slow convergence for at least a number of iterations depending on how indefinite the matrix is).

In practice, GMRES(m) is often used instead of full GMRES. My small examples already hint at some of the problems associated with the selection of m. If a value of m is selected that is too small to overcome a phase of relatively slow convergence (a phase with a bulge in the reduction factors) then it is plausible that we will never see the faster convergence of full GMRES. This is most obvious for the example considered in Exercise 6.11. For that example we will never see convergence for $m < n$. In practice, we will rarely see such extreme situations. However, the choice for m remains tricky and it may occur that a specific value of m gives relatively slow convergence, whereas the value $m + 1$ would result in much faster convergence. As a rule of thumb it seems to be good practice to select m so that the field of values $\mathcal{F}(H_{m,m})$ coincides more or less with $\mathcal{F}(H_{m+1,m+1})$. If it is to be expected that the (preconditioned) matrix is not too far from a normal matrix, then the convex hull of the eigenvalues of $H_{m,m}$ may be taken to approximate $\mathcal{F}(H_{m,m})$.

It should be noted here that it is not always efficient to enlarge m. Not only is GMRES increasingly expensive per iteration, it may also be the case that GMRES($m - 1$) converges faster than GMRES(m) in terms of matrix vector multiplications. For an example of this, see [77, Chapter 4].

6.4 MINRES

When A is symmetric, then the matrix $H_{i+1,i}$ reduces to a tridiagonal matrix $T_{i+1,i}$. This property can be exploited to obtain short recurrence relations. As with GMRES, we look for an

$$x_i \in \{r_0, Ar_0, \ldots, A^{i-1}r_0\}, \quad x_i = R_i\bar{y}$$

$$\|Ax_i - b\|_2 = \|AR_i\bar{y} - b\|_2$$

$$= \|R_{i+1}T_{i+1,i}y - b\|_2,$$

such that this residual norm is minimal. Now we exploit the fact that $R_{i+1}D_{i+1}^{-1}$, with

$$D_{i+1} = diag(\|r_0\|_2, \|r_1\|_2, \ldots, \|r_i\|_2),$$

is an orthonormal transformation with respect to the current Krylov subspace:

$$\|Ax_i - b\|_2 = \|D_{i+1}T_{i+1,i}y - \|r_0\|_2 e_1\|_2$$

and this final expression can simply be seen as a minimum norm least squares problem.

The element in the $(i+1, i)$ position of $T_{i+1,i}$ can be transformed to zero by a simple Givens rotation and the resulting upper tridiagonal system (the other subdiagonal elements being removed in previous iteration steps) can simply be solved, as I will now show.

The effect of the Givens rotations is that $T_{i+1,i}$ is decomposed in QR-form:

$$T_{i+1,i} = Q_{i+1,i} R_{i,i},$$

in which the orthogonal matrix $Q_{i+1,i}$ is the product of the Givens rotations and $R_{i,i}$ is an upper triangular matrix with three nonzero diagonals. We can exploit the banded structure of $R_{i,i}$, for the computation of x_i.

The solution of $T_{i+1,i}y = ||r_0||_2 e_1$ can be written as

$$y = R_{i,i}^{-1} Q_{i+1,i}^T ||r_0||_2 e_1,$$

so that the solution x_i is obtained as

$$x_i = (V_i R_{i,i}^{-1})(Q_{i+1,i}^T \|r_0\|_2 e_1).$$

We first compute the matrix $W_k = V_k R_k^{-1}$, and it is easy to see that the last column of W_k is obtained from the last 3 columns of V_k. The vector $z_i \equiv Q_{i+1,i}^T \|r_0\|_2 e_1$ can also be updated by a short recurrence, since z_i follows from a simple Givens rotation on the last two coordinates of z_{i-1}. These two coupled recurrence relations lead to MINRES [153].

This leads to the algorithmic form of MINRES in Figure 6.9, which has been inspired by a MATLAB routine published in [82].

MINRES is attractive when the symmetric matrix A is symmetric and indefinite. In the positive definite case, CG is the preferred method. For a symmetric positive definite preconditioner of the form LL^T, the MINRES algorithm can be applied to the explicitly preconditioned system

$$L^{-1}AL^{-T}\widetilde{x} = L^{-1}b, \text{ with } x = L^{-T}\widetilde{x}.$$

Explicit inversions with L and L^T can be avoided, all we have to do is solve linear systems with these matrices. We cannot, without risk, apply MINRES to $K^{-1}Ax = K^{-1}b$, for symmetric K and A, when the matrix $K^{-1}A$ is unsymmetric. It does not help to replace the inner product by the bilinear form (x, Ky), with respect to which the matrix $K^{-1}A$ is symmetric, since this bilinear form does not define an inner product if K is not positive definite. The

Compute $v_1 = b - Ax_0$ for some initial guess x_0
$\beta_1 = ||v_1||_2$; $\eta = \beta_1$;
$\gamma_1 = \gamma_0 = 1$; $\sigma_1 = \sigma_0 = 0$;
$v_0 = 0$; $w_0 = w_{-1} = 0$;
for $i = 1, 2, \ldots.$
The Lanczos recurrence:
$v_i = \frac{1}{\beta_i} v_i$; $\alpha_i = v_i^T A v_i$;
$v_{i+1} = A v_i - \alpha_i v_i - \beta_i v_{i-1}$
$\beta_{i+1} = ||v_{i+1}||_2$
QR part:
Old Givens rotations on new column of T:
$\delta = \gamma_i \alpha_i - \gamma_{i-1} \sigma_i \beta_i$; $\rho_1 = \sqrt{\delta^2 + \beta_{i+1}^2}$
$\rho_2 = \sigma_i \alpha_i + \gamma_{i-1} \gamma_i \beta_i$; $\rho_3 = \sigma_{i-1} \beta_i$
New Givens rotation for subdiag element:
$\gamma_{i+1} = \delta / \rho_1$; $\sigma_{i+1} = \beta_{i+1} / \rho_1$
Update of solution (with $W_i = V_i R_{i,i}^{-1}$)
$w_i = (v_i - \rho_3 w_{i-2} - \rho_2 w_{i-1}) / \rho_1$
$x_i = x_{i-1} + \gamma_{i+1} \eta w_i$
$||r_i||_2 = |\sigma_{i+1}| ||r_{i-1}||_2$
check convergence; continue if necessary
$\eta = -\sigma_{i+1} \eta$
end

Figure 6.9. The unpreconditioned MINRES algorithm.

construction of effective preconditioners for symmetric indefinite A is largely an open problem.

With respect to parallelism, or other implementation aspects, MINRES can be treated as CG. Note that most of the variables in MINRES may overwrite old ones that are obsolete.

The use of the 3-term recurrence relation for the columns of W_i makes MINRES very vulnerable to rounding errors, as has been shown in [179]. It has been shown that rounding errors are propagated to the approximate solution with a factor proportional to the square of the condition number of A, whereas in GMRES these errors depend only on the condition number itself. Therefore, we should be careful with MINRES for ill-conditioned systems. If storage is no problem then GMRES should be preferred for ill-conditioned systems; if storage is a problem then we might consider the usage of SYMMLQ [153]. SYMMLQ, however, may converge a good deal slower than MINRES for ill-conditioned systems. For more details on this see Section 8.3.

6.5 Rank-one updates for the matrix splitting

Iterative methods can be derived from a splitting of the matrix, and we have used the very simple splitting $A = I - R$, with $R = I - A$, in order to derive the projection type methods. In [73] updating the matrix splitting with information obtained in the iteration process is suggested. I will give the flavour of this method here since it turns out that it has an interesting relation with GMRES. This relation is exploited in [207] for the construction of new classes of GMRES-like methods that can be used as cheap alternatives for the increasingly expensive full GMRES method.

If we assume that the matrix splitting in the k-th iteration step is given by $A = H_k^{-1} - R_k$, we obtain the iteration formula

$$x_k = x_{k-1} + H_k r_{k-1} \quad \text{with} \quad r_k = b - Ax_k.$$

The idea now is to construct H_k by a suitable rank-one update to H_{k-1}:

$$H_k = H_{k-1} + u_{k-1} v_{k-1}^T,$$

which leads to

$$x_k = x_{k-1} + (H_{k-1} + u_{k-1} v_{k-1}^T) r_{k-1} \tag{6.12}$$

or

$$\begin{aligned} r_k &= r_{k-1} - A(H_{k-1} + u_{k-1} v_{k-1}^T) r_{k-1} \\ &= (I - AH_{k-1}) r_{k-1} - A u_{k-1} v_{k-1}^T r_{k-1} \\ &= (I - AH_{k-1}) r_{k-1} - \mu_{k-1} A u_{k-1}. \end{aligned} \tag{6.13}$$

The optimal choice for the update would have been to select u_{k-1} such that

$$\mu_{k-1} A u_{k-1} = (I - AH_{k-1}) r_{k-1},$$

or

$$\mu_{k-1} u_{k-1} = A^{-1} (I - AH_{k-1}) r_{k-1}.$$

However, A^{-1} is unknown and the best approximation we have for it is H_{k-1}. This leads to the choice

$$\bar{u}_{k-1} = H_{k-1} (I - AH_{k-1}) r_{k-1}. \tag{6.14}$$

The constant μ_{k-1} is chosen such that $\|r_k\|_2$ is minimal as a function of μ_{k-1}. This leads to

$$\mu_{k-1} = \frac{1}{\|A\bar{u}_{k-1}\|_2^2}(A\bar{u}_{k-1})^T(I - AH_{k-1})r_{k-1}.$$

Since v_{k-1} has to be chosen such that $\mu_{k-1} = v_{k-1}^T r_{k-1}$, we have the following obvious choice for it

$$\bar{v}_{k-1} = \frac{1}{\|A\bar{u}_{k-1}\|_2^2}(I - AH_{k-1})^T A\bar{u}_{k-1} \tag{6.15}$$

(note that from the minimization property we have that $r_k \perp A\bar{u}_{k-1}$).

In principle the implementation of the method is quite straightforward, but note that the computation of r_{k-1}, $\bar{u}_{k-1}$, and $\bar{v}_{k-1}$ costs 4 matrix vector multiplications with A (and also some with H_{k-1}). This would make the method too expensive to be of practical interest. Also the updated splitting is most likely a dense matrix if we carry out the updates explicitly. I will now show, still following the lines set forth in [73], that there are orthogonality properties, following from the minimization step, by which the method can be implemented much more efficiently.

We define

(1) $c_k = \frac{1}{\|A\bar{u}_k\|_2}A\bar{u}_k$ (note that $r_{k+1} \perp c_k$),
(2) $E_k = I - AH_k$.

From (6.13) we have that $r_k = E_k r_{k-1}$, and from (6.14):

$$A\bar{u}_k = AH_k E_k r_k = \alpha_k c_k$$

or

$$c_k = \frac{1}{\alpha_k}(I - E_k)E_k r_k = \frac{1}{\alpha_k}E_k(I - E_k)r_k. \tag{6.16}$$

Furthermore:

$$E_k = I - AH_k = I - AH_{k-1} - A\bar{u}_{k-1}\bar{v}_{k-1}^T$$

$$(6.14) \Rightarrow \quad = I - AH_{k-1} - A\bar{u}_{k-1}(A\bar{u}_{k-1})^T(I - AH_{k-1})\frac{1}{\|A\bar{u}_k\|_2^2}$$

$$= (I - c_{k-1}c_{k-1}^T)E_{k-1}$$

$$= \prod_{i=0}^{k-1}(I - c_i c_i^T)E_0. \tag{6.17}$$

We see that the operator E_k has the following effect on a vector. The vector is multiplied by E_0 and then orthogonalized with respect to $c_0, \ldots, c_{k-1}$. Now we have from (6.16) that

$$c_k = \frac{1}{\alpha_k} E_k y_k,$$

and hence

$$c_k \perp c_0, \ldots, c_{k-1}. \tag{6.18}$$

A consequence from (6.18) is that

$$\prod_{j=0}^{k-1} (I - c_j c_j^T) = I - \sum_{j=0}^{k-1} c_j c_j^T = I - P_{k-1}$$

and therefore

$$P_k = \sum_{j=0}^{k} c_j c_j^T. \tag{6.19}$$

The actual implementation is based on the above properties. Given r_k we compute r_{k+1} as follows (and we update x_k in the corresponding way):

$$r_{k+1} = E_{k+1} r_k.$$

With $\xi^{(0)} = E_0 r_k$ we first compute (with the c_j from previous steps):

$$E_k r_k = \xi^{(k)} \equiv \left(I - \sum_{j=0}^{k-1} c_j c_j^T \right) \xi^{(0)} = \prod_{j=0}^{k-1} (I - c_j c_j^T) \xi^{(0)}.$$

The expression with $\sum$ leads to a Gram–Schmidt formulation, the expression with $\prod$ leads to the Modified Gram-Schmidt variant. The computed updates $-c_j^T \xi^{(0)} c_j$ for r_{k+1} correspond to updates

$$c_j^T \xi^{(0)} A^{-1} c_j = c_j^T \xi^{(0)} u_j / \|Au_j\|_2$$

for x_{j+1}. These updates are in the scheme, given below, represented by η.

From (6.14) we know that

$$\bar{u}_k = H_k E_k r_k = H_k \xi^{(k)}.$$

Now we have to make $A\bar{u}_k \sim c_k$ orthogonal with respect to $c_0, \ldots, c_{k-1}$, and to update $\bar{u}_k$ accordingly. Once we have done that we can do the final update step to make H_{k+1}, and we can update both x_k and r_k by the corrections following from including c_k. The orthogonalization step can be carried out easily as follows. Define $c_k^{(k)} \equiv \alpha_k c_k = AH_k E_k r_k = (I - E_k)E_k r_k$ (see (6.16)) $= (I - E_0 + P_{k-1}E_0)\xi^{(k)}$ (see (6.17)) $= AH_0\xi^{(k)} + P_{k-1}(I - AH_0)\xi^{(k)} = c_k^{(0)} + P_{k-1}\xi^{(k)} - P_{k-1}c_k^{(0)}$. Note that the second term vanishes since $\xi^{(k)} \perp c_0, \ldots, c_{k-1}$.

The resulting scheme for the k-th iteration step becomes:

(1) $\xi^{(0)} = (I - AH_0)r_k;\ \eta^{(0)} = H_0 r_k;$
  for $i = 0, \ldots, k-1$ do
    $\alpha_i = c_i^T\xi^{(i)};\ \xi^{(i+1)} = \xi^{(i)} - \alpha_i c_i;\ \eta^{(i+1)} = \eta^{(i)} + \alpha_i u_i;$
(2) $u_k^{(0)} = H_0\xi^{(k)};\ c_k^{(0)} = Au_k^{(0)};$
  for $i = 0, \ldots, k-1$ do
    $\beta_i = -c_i^T c_k^{(i)};\ c_k^{(i+1)} = c_k^{(i)} + \beta_i c_i;\ u_k^{(i+1)} = u_k^{(i)} + \beta_i u_i;$
  $c_k = c_k^{(k)}/\|c_k^{(k)}\|_2;\ u_k = u_k^{(k)}/\|c_k^{(k)}\|_2;$
(3) $x_{k+1} = x_k + \eta^{(k)} + u_k c_k^T\xi^{(k)};$
  $r_{k+1} = (I - c_k c_k^T)\xi^{(k)};$

Remarks

(1) The above scheme is a Modified Gram–Schmidt variant, given in [207], of the original scheme in [73].
(2) If we keep H_0 fixed, i.e., $H_0 = I$, then the method is not scaling invariant (the results for $\rho Ax = \rho b$ depend on ρ). In [207] a scaling invariant method is suggested.
(3) Note that in the above implementation we have 'only' two matrix vector products per iteration step. In [207] it is shown that in many cases we may also expect convergence comparable to GMRES in half the number of iteration steps.
(4) A different choice for $\bar{u}_{k-1}$ does not change the formulas for $\bar{v}_{k-1}$ and E_{k-1}. For each different choice we can derive schemes similar to the one above.
(5) From (6.13) we have

$$r_k = r_{k-1} - AH_{k-1}r_{k-1} - \mu_{k-1}Au_{k-1}.$$

In view of the previous remark we might also make the different choice $\bar{u}_{k-1} = H_{k-1}r_{k-1}$. With this choice, we obtain a variant which is algebraically identical to GMRES (for a proof of this see [207]). This GMRES variant is obtained by the following changes in the previous scheme: Take $H_0 = 0$ (note that in this case we have that $E_{k-1}r_{k-1} = r_{k-1}$, and hence we may skip step (1) of the above algorithm), and set $\xi^{(k)} = r_k$, $\eta^{(k)} = 0$. In step (2) start with $u_k^{(0)} = \xi^{(k)}$.

The result is a different formulation of GMRES in which we can obtain explicit formulas for the updated preconditioner (i.e., the inverse of A is approximated increasingly well): The update for H_k is $\bar{u}_k c_k^T E_k$ and the sum of these updates gives an approximation for A^{-1}.

(6) Also in this GMRES-variant we are still free to select a slightly different u_k. Remember that the leading factor H_{k-1} in (6.14) was introduced as an approximation for the actually desired A^{-1}. With $\bar{u}_{k-1} = A^{-1} r_{k-1}$, we would have that $r_k = E_{k-1} r_{k-1} - \mu_{k-1} r_{k-1} = 0$ for the minimizing μ_{k-1}. We could take other approximations for the inverse (with respect to the given residual r_{k-1}), e.g., the result vector y obtained by a few steps of GMRES applied to $Ay = r_{k-1}$. This leads to the so-called GMRESR family of nested methods (for details see [207]), see also Section 6.6. A similar algorithm, named FGMRES, has been derived independently by Saad [166]. In FGMRES the search directions are preconditioned, whereas in GMRESR the residuals are preconditioned. This gives GMRESR direct control over the reduction in norm of the residual. As a result GMRESR can be made robust while FGMRES may suffer from breakdown. A further disadvantage of the FGMRES formulation is that this method cannot be truncated, or selectively orthogonalized, as GMRESR can be.

In [14] a generalized conjugate gradient method is proposed, a variant of which produces in exact arithmetic identical results as the proper variant of GMRESR, though at higher computational costs and with a classical Gram–Schmidt orthogonalization process instead of the modified process in GMRESR.

6.6 GMRESR and GMRES⋆

In [207] it has been shown that the GMRES method can be effectively combined (or rather preconditioned) with other iterative schemes. The iteration steps of GMRES (or GCR) are called outer iteration steps, while the iteration steps of the preconditioning iterative method are referred to as inner iterations. The combined method is called GMRES⋆, where ⋆ stands for any given iterative scheme; in the case of GMRES as the inner iteration method, the combined scheme is called GMRESR [207].

In exact arithmetic GMRES⋆ is very close to the Generalized Conjugate Gradient method [14]; GMRES⋆, however, leads to a more efficient computational scheme.

```
x_0 is an initial guess; r_0 = b - A x_0;
for i = 0, 1, 2, 3, ....
    Let z^(m) be the approximate solution of A z = r_i
    obtained after m steps of an iterative method.
    c = A z^(m) (often available from the iterative method)
    for k = 0, ..., i - 1
        α = (c_k, c)
        c = c - α c_k
        z^(m) = z^(m) - α u_k
    end
    c_i = c/||c||_2; u_i = z^(m)/||c||_2
    x_{i+1} = x_i + (c_i, r_i) u_i
    r_{i+1} = r_i - (c_i, r_i) c_i
    if x_{i+1} is accurate enough then quit
end
```

Figure 6.10. The GMRES⋆ algorithm.

The GMRES⋆ algorithm can be described by the computational scheme in Figure 6.10.

A sufficient condition to avoid breakdown in this method ($\|c\|_2 = 0$) is that the norm of the residual at the end of an inner iteration is smaller than the right-hand residual: $\|Az^{(m)} - r_i\|_2 < \|r_i\|_2$. This can easily be controlled during the inner iteration process. If stagnation occurs, i.e. no progress at all is made in the inner iteration, then van der Vorst and Vuik [207] suggest doing one (or more) steps of the LSQR method, which guarantees a reduction (but this reduction is often only small).

When memory space is a limiting factor or when the computational costs per iteration become too high, we can simply *truncate* the algorithm (instead of restarting as in GMRES(m)). If we wish only to retain the last m vectors c_i and u_i, the truncation is effected by replacing the **for** k loop in Figure 6.10 by

$$\textbf{for } k = \max(0, i - m), \ldots, i - 1$$

and of course, we have to adapt the remaining part of the algorithm so that only the last m vectors are kept in memory.

Exercise 6.12. *Modify the algorithm in Figure 6.10 so that we obtain the truncated variant and so that only the last m vectors u_i and c_i are kept in memory. Compare this truncated algorithm with GMRES(m) for the examples in Exercise 6.3.*

The idea behind the nested iteration scheme in GMRESR is that we explore parts of high-dimensional Krylov subspaces, hopefully localizing the same approximate solution that full GMRES would find over the entire subspace, but now at much lower computational costs. The alternatives for the inner iteration could be either one cycle of GMRES(m), since then we also have locally an optimal method, or some other iteration scheme, such as Bi-CGSTAB. As has been shown by van der Vorst [202], there are a number of different situations where we may expect stagnation or slow convergence for GMRES(m). In such cases it does not seem wise to use this method.

On the other hand it may also seem questionable whether a method like Bi-CGSTAB should lead to success in the inner iteration. This method does not satisfy a useful global minimization property and a large part of its effectiveness comes from the underlying Bi-CG algorithm, which is based on bi-orthogonality relations. This means that for each outer iteration the inner iteration process again has to build a bi-orthogonality relation. It has been shown for the related Conjugate Gradients method that the orthogonality relations are determined largely by the distribution of the weights at the lower end of the spectrum and on the isolated eigenvalues at the upper end of the spectrum [193]. By the nature of these kinds of Krylov process, the largest eigenvalues and their corresponding eigenvector components quickly enter the process after each restart, and hence it may be expected that much of the work is taken up in rediscovering the same eigenvector components in the error over and over again, whereas these components may already be so small that a further reduction in those directions in the outer iteration is a waste of time, since it hardly contributes to a smaller norm of the residual. This heuristic way of reasoning may explain in part our rather disappointing experiences with Bi-CGSTAB as the inner iteration process for GMRES⋆.

De Sturler and Fokkema [54] propose that the outer search directions are explicitly prevented from being reinvestigated again in the inner process. This is done by keeping the Krylov subspace that is built in the inner iteration orthogonal with respect to the Krylov basis vectors generated in the outer iteration. The procedure works as follows.

In the outer iteration process the vectors $c_0, \ldots, c_{i-1}$ build an orthogonal basis for the Krylov subspace. Let C_i be the n by i matrix with columns $c_0, \ldots, c_{i-1}$. Then the inner iteration process at outer iteration i is carried out with the operator A_i instead of A, and A_i is defined as

$$A_i = (I - C_i C_i^T)A. \tag{6.20}$$

It is easily verified that $A_i z \perp c_0, \ldots, c_{i-1}$ for all z, so that the inner iteration process takes place in a subspace orthogonal to these vectors. The additional

7

Bi-Conjugate Gradients

7.1 Derivation of the method

For the computation of an approximation x_i for the solution x of $Ax = b$, $x_i \in \mathcal{K}^i(A, r_0)$, $x_0 = 0$, with unsymmetric A, we can start again from the Lanczos relations (similarly to the symmetric case, cf. Chapter 5):

$$AV_i = V_{i+1}T_{i+1,i}, \tag{7.1}$$

but here we use the matrix $W_i = [w_1, w_2, \ldots, w_i]$ for the projection of the system

$$W_i^T(b - Ax_i) = 0,$$

or

$$W_i^T AV_i y - W_i^T b = 0.$$

Using (7.1), we find that y_i satisfies

$$T_{i,i}y = \|r_0\|_2 e_1,$$

and $x_i = V_i y$. The resulting method is known as the Bi-Lanczos method [130].

We have assumed that $d_i \neq 0$, that is $w_i^T v_i \neq 0$. The generation of the bi-orthogonal basis breaks down if for some i the value of $w_i^T v_i = 0$, this is referred to in literature as a *serious breakdown*. Likewise, when $w_i^T v_i \approx 0$, we have a near-breakdown. The way to get around this difficulty is the so-called *look-ahead* strategy, which takes a number of successive basis vectors for the

Krylov subspace together and makes them blockwise bi-orthogonal. This has been worked out in detail in [156, 89, 90, 91].

Another way to avoid breakdown is to restart as soon as a diagonal element becomes small. Of course, this strategy looks surprisingly simple, but it should be realised that at a restart the Krylov subspace, which has been built up so far, is thrown away, and this destroys the possibility of faster (i.e., superlinear) convergence.

We can try to construct an LU-decomposition, without pivoting, of $T_{i,i}$. If this decomposition exists, then, similarly to CG, it can be updated from iteration to iteration and this leads to a recursive update of the solution vector, which avoids saving all intermediate r and w vectors. This variant of Bi-Lanczos is usually called Bi-Conjugate Gradients, or Bi-CG for short [83]. In Bi-CG, the d_i are chosen such that $v_i = r_{i-1}$, similarly to CG.

Of course we cannot in general be certain that an LU decomposition (without pivoting) of the tridiagonal matrix $T_{i,i}$ exists, and this may also lead to a breakdown (a breakdown of the *second kind*) of the Bi-CG algorithm. Note that this breakdown can be avoided in the Bi-Lanczos formulation of the iterative solution scheme, e.g., by making an LU-decomposition with 2 by 2 block diagonal elements [17]. It is also avoided in the QMR approach (see Section 7.3).

Note that for symmetric matrices Bi-Lanczos generates the same solution as Lanczos, provided that $w_1 = r_0$, and under the same condition Bi-CG delivers the same iterands as CG for positive definite matrices. However, the bi-orthogonal variants do so at the cost of two matrix vector operations per iteration step. For a computational scheme for Bi-CG, without provisions for breakdown, see Figure 7.1.

The scheme in Figure 7.1 may be used for a computer implementation of the Bi-CG method. In the scheme the equation $Ax = b$ is solved with a suitable preconditioner K.

As with conjugate gradients, the coefficients α_j and β_j, $j = 0, \ldots, i-1$ build the matrix T_i, as given in formula (5.10). This matrix is, for Bi-CG, not generally similar to a symmetric matrix. Its eigenvalues can be viewed as Petrov–Galerkin approximations, with respect to the spaces $\{\widetilde{r}_j\}$ and $\{r_j\}$, of eigenvalues of A. For increasing values of i they tend to converge to eigenvalues of A. The convergence patterns, however, may be much more complicated and irregular than in the symmetric case.

```
x_0 is an initial guess, r_0 = b − Ax_0
Choose r̃_0, for example r̃_0 = r_0
for i = 1, 2, ....
    Solve K w_{i−1} = r_{i−1}
    Solve K^T w̃_{i−1} = r̃_{i−1}
    ρ_{i−1} = w̃_{i−1}^T w_{i−1}
    if ρ_{i−1} = 0 method fails
    if i = 1
        p_i = w_{i−1}
        p̃_i = w̃_{i−1}
    else
        β_{i−1} = ρ_{i−1}/ρ_{i−2}
        p_i = w_{i−1} + β_{i−1} p_{i−1}
        p̃_i = w̃_{i−1} + β_{i−1} p̃_{i−1}
    endif
    z_i = A p_i
    z̃_i = A^T p̃_i
    α_i = ρ_{i−1}/p̃_i^T z_i)
    x_i = x_{i−1} + α_i p_i
    r_i = r_{i−1} − α_i z_i
    r̃_i = r̃_{i−1} − α_i z̃_i
    if x_i is accurate enough then quit
end
```

Figure 7.1. Bi-CG algorithm.

7.2 Another derivation of Bi-CG

An alternative way to derive Bi-CG comes from considering the following symmetric linear system:

$$\begin{pmatrix} 0 & A \\ A^T & 0 \end{pmatrix} \begin{pmatrix} \widehat{x} \\ x \end{pmatrix} = \begin{pmatrix} b \\ \widehat{b} \end{pmatrix}, \quad \text{or } B\widetilde{x} = \widetilde{b},$$

for some suitable vector $\widehat{b}$.

If we select $\widehat{b} = 0$ and apply the CG-scheme to this system, then again we obtain LSQR. However, if we select $\widehat{b} \neq 0$ and apply the CG scheme with the preconditioner

$$\begin{pmatrix} 0 & I \\ I & 0 \end{pmatrix},$$

in the way shown in Section 5.2, then we immediately obtain the unpreconditioned Bi-CG scheme for the system $Ax = b$. Note that the CG-scheme can be applied since $K^{-1}B$ is symmetric (but not positive definite) with respect to the bilinear form

$$[p, q] \equiv (p, Kq),$$

which is not a proper inner product. Hence, this formulation clearly reveals the two principal weaknesses of Bi-CG (i.e., the causes for breakdown situations). Note that if we restrict ourselves to vectors

$$p = \begin{pmatrix} p_1 \\ p_1 \end{pmatrix},$$

then $[p, q]$ defines a proper inner product. This situation arises for the Krylov subspace that is created for B and $\widetilde{b}$ if $A = A^T$ and $\widehat{b} = b$. If, in addition, A is positive definite then $K^{-1}B$ is positive definite symmetric with respect to the generated Krylov subspace, and we obtain the CG-scheme (as expected). More generally, the choice

$$K = \begin{pmatrix} 0 & K_1 \\ K_1^T & 0 \end{pmatrix},$$

where K_1 is a suitable preconditioner for A, leads to the preconditioned version of the Bi-CG scheme, as given in Figure 7.1.

The above presentation of Bi-CG was inspired by a closely related presentation of Bi-CG in [119]. The latter paper gives a rather untractable reference for the choice of the system $B\widetilde{x} = \widetilde{b}$ and the preconditioner

$$\begin{pmatrix} 0 & I \\ I & 0 \end{pmatrix}$$

to [120].

7.3 QMR

The QMR method [91] relates to Bi-CG similarly to the way in which MINRES relates to CG. We start with the recurrence relations for the v_j:

$$AV_i = V_{i+1}T_{i+1,i}.$$

We would like to identify the x_i, with $x_i \in K^i(A; r_0)$, or $x_i = V_i y$, for which

$$\|b - Ax_i\|_2 = \|b - AV_i y\|_2 = \|b - V_{i+1}T_{i+1,i}y\|_2$$

is minimal, but the problem is that V_{i+1} is not orthogonal. However, we pretend that the columns of V_{i+1} are orthogonal. Then

$$\|b - Ax_i\|_2 = \|V_{i+1}(\|r_0\|_2 e_1 - T_{i+1,i}y)\|_2 = \|(\|r_0\|_2 e_1 - T_{i+1,i}y)\|_2,$$

and in [91] solving the projected miniminum norm least squares problem $\|(\|r_0\|_2 e_1 - T_{i+1,i}y)\|_2$ is suggested. The minimum value of this norm is called the quasi residual and will be denoted by $\|r_i^Q\|_2$.

Since, in general, the columns of V_{i+1} are not orthogonal, the computed $x_i = V_i y$ does not solve the minimum residual problem, and therefore this approach is referred to as a quasi-minimum residual approach called QMR [91]. It can be shown that the norm of the residual r_i^{QMR} of QMR can be bounded in terms of the quasi residual

$$\|r_i^{QMR}\|_2 \le \sqrt{i+1}\,\|r_i^Q\|_2.$$

The above sketched approach leads to the simplest form of the QMR method. A more general form arises if the least squares problem is replaced by a weighted least squares problem [91]. No strategies are yet known for optimal weights.

In [91] the QMR method is carried out on top of a look-ahead variant of the bi-orthogonal Lanczos method, which makes the method more robust. Experiments indicate that although QMR has a much smoother convergence behaviour than Bi-CG, it is not essentially faster than Bi-CG. This is confirmed explicitly by the following relation for the Bi-CG residual r_k^B and the quasi residual r_k^Q (in exact arithmetic):

$$\|r_k^B\|_2 = \frac{\|r_k^Q\|_2}{\sqrt{1 - (\|r_k^Q\|_2/\|r_{k-1}^Q\|_2)^2}}, \tag{7.2}$$

see [49, Theorem 4.1]. This relation, which is similar to the relation for GMRES and FOM, shows that when QMR gives a significant reduction at step k, then Bi-CG and QMR have arrived at residuals of about the same norm (provided, of course, that the same set of starting vectors has been used).

It is tempting to compare QMR with GMRES, but this is difficult. GMRES really minimizes the 2-norm of the residual, but at the cost of increasing the work of keeping all residuals orthogonal and increasing demands for memory space. QMR does not minimize this norm, but often it has a convergence comparable to GMRES, at a cost of twice the amount of matrix vector products per iteration step. However, the generation of the basis vectors in QMR is relatively cheap and the memory requirements are limited and modest. The relation (7.2) expresses that at a significant local error reduction of QMR, Bi-CG and QMR have arrived almost at the same residual vector (similarly to GMRES and FOM). However, QMR is preferred to Bi-CG in all cases because of its much smoother convergence behaviour, and also because QMR removes one breakdown condition (even when implemented without look-ahead). Several variants of QMR, or rather Bi-CG, have been proposed that increase the effectiveness of this class of methods in certain circumstances [85]. See Section 7.5 for a variant that is suitable for complex symmetric systems. In Figure 7.2 we present the simplest form of QMR, that is without look-ahead [20], for the solution of $Ax = b$, for real A and with preconditioner $M = M_1 M_2$.

Zhou and Walker [226] have shown that the QMR approach can also be followed for other methods, such as CGS and Bi-CGSTAB. The main idea is that in these methods the approximate solution is updated as

$$x_{i+1} = x_i + \alpha_i p_i,$$

and the corresponding residual is updated as

$$r_{i+1} = r_i - \alpha_i A p_i.$$

This means that $AP_i = W_i R_{i+1}$, with W_i a lower bidiagonal matrix. The x_i are combinations of the p_i, so that we can try to find the combination $P_i y_i$ for which $\|b - AP_i y_i\|_2$ is minimal. If we insert the expression for AP_i, and ignore the fact that the r_i are not orthogonal, then we can minimize the norm of the residual in a quasi-minimum least squares sense, similarly to QMR.

Exercise 7.1. *The methods Bi-CG and QMR are suitable for general unsymmetric nonsingular linear systems. Construct test cases with known spectra. How is working avoided with dense matrices? Try to get an impression of systems for which Bi-CG and QMR work well. Is it possible to create linear systems for which Bi-CG nearly breaks down at step i? Hint: compute the Ritz values at step i and consider a shift for the matrix of the system.*

Exercise 7.2. *In the methods Bi-CG and QMR, the iteration coefficients build a tridiagonal matrix T. Bi-CG is based on the i by i part, QMR works with the $i+1$ by i part. The difference is thus only in one element. Could that explain effects in the convergence behaviour of both methods? Have the eigenvalues of*

x_0 is an initial guess, $r_0 = b - Ax_0$, $\widetilde{v} = r_0$
Choose $\widetilde{w}$, for example $\widetilde{w} = r_0$
Solve $M_1 y = \widetilde{v}$, $\rho_1 = \|y\|_2$, solve $M_2^T z = \widetilde{w}$, $\xi_1 = \|z\|_2$
for $i = 1, 2, \ldots.$
  if $\rho_i = 0$ **or** $\xi_i = 0$ method fails
  $v_i = \widetilde{v}/\rho_i$, $y = y/\rho_i$, $w_i = \widetilde{w}/\xi_i$, $z = z/\xi_i$
  $\delta_i = z^T y$; **if** $\delta_i = 0$ method fails
  Solve $M_2 \widetilde{y} = y$, solve $M_1^T \widetilde{z} = z$
  if $i = 1$
    $p_1 = \widetilde{y}$, $q_1 = \widetilde{z}$
  else
    $p_i = \widetilde{y} - (\xi_i \delta_i / \epsilon_{i-1}) p_{i-1}$
    $q_i = \widetilde{z} - (\rho_i \delta_i / \epsilon_{i-1}) q_{i-1}$
  endif
  $\widetilde{p} = A p_i$
  $\epsilon_i = q_i^T \widetilde{p}$, **if** $\epsilon_i = 0$ method fails
  $\beta_i = \epsilon_i / \delta_i$, $\widetilde{v} = \widetilde{p} - \beta_i v_i$
  Solve $M_1 y = \widetilde{v}$, $\rho_{i+1} = \|y\|_2$
  $\widetilde{w} = A^T q_i - \beta_i w_i$
  Solve $M_2^T z = \widetilde{w}$, $\xi_{i+1} = \|z\|_2$
  $\theta_i = \rho_{i+1} / (\gamma_{i-1} |\beta_i|)$, $\gamma_i = 1/\sqrt{1 + \theta_i^2}$, **if** $\gamma_i = 0$ method fails
  $\eta_i = -\eta_{i-1} \rho_i \gamma_i^2 / (\beta_i \gamma_{i-1}^2)$
  if $i = 1$
    $d_1 = \eta_1 p_1$, $s_1 = \eta_1 \widetilde{p}$
  else
    $d_i = \eta_i p_i + (\theta_{i-1} \gamma_i)^2 d_{i-1}$
    $s_i = \eta_i \widetilde{p} + (\theta_{i-1} \gamma_i)^2 s_{i-1}$
  endif
  $x_i = x_{i-1} + d_i$, $r_i = r_{i-1} - s_i$
  if x_i is accurate enough **then** quit
end

Figure 7.2. QMR without look-ahead and preconditioner $M = M_1 M_2$.

$T_{i,i}$ any relation with those of the system matrix? Is there an explanation for such a relation?

In Sections 6.1.2 and 6.6 we have seen variants of GMRES, namely FGMRES and GMRESR that permit variable preconditioning. Such variable preconditioning may be advantageous in situations where the preconditioning is different because of parallel processing or when the preconditioning operator is defined with, for instance, a sweep of multigrid with iterative smoothing. In all these variable preconditioning methods we need to store two different complete bases, and the question is whether flexible variants of short-term recurrence methods exist. Unfortunately, this is not the case, because the global (bi-)orthogonality requires fixed operators. The so-called Flexible QMR (FQMR) method [185] seems to contradict that statement. However, as is shown in [185, Theorem 2.2] there is only a very local bi-orthogonality between the two sets of bases in FQMR. That is, under some restrictive assumptions, it can be proved that $v_i^T w_{i-1} = 0$ and $w_i^T v_{i-1} = 0$. One of the assumptions is that the preconditioner K_i is the same for A and A^T in the i-th iteration step. Nevertheless, even when this assumption is violated, numerical experiments show that FQMR may converge rather quickly when the preconditioning operation is carried out by a preconditioned QMR process again. It is not quite clear yet under what circumstances it can be advantageous to use FGMRES. In [185] the argument of higher eventual accuracy of the approximated solution is given, based upon experimental observations. However, it might be that the differences in accuracy between QMR and FQMR in these experiments stem from the updating procedure and it may be that they disappear when using *reliable updating*, see Section 8.1.

7.4 CGS

For the derivation of the CGS method we exploit the fact that the vectors belong to some Krylov subspace and, in particular, we write these vectors in polynomial form. Our derivation is based on the Bi-CG algorithm. Note that the bi-conjugate gradient residual vector can be written as r_j $(= \rho_j v_j) = P_j(A) r_0$, and, similarly, the so-called shadow residual $\widetilde{r}_j$ $(= \rho_j w_j)$ can be written as $\widetilde{r}_j = P_j(A^T)\widetilde{r}_0$. Because of the bi-orthogonality relation we have that

$$\begin{aligned}(r_j, \widetilde{r}_i) &= (P_j(A) r_0, P_i(A^T)\widetilde{r}_0) \\ &= (P_i(A) P_j(A) r_0, \widetilde{r}_0) = 0,\end{aligned}$$

for $i < j$. The iteration parameters for bi-conjugate gradients are computed from similar inner products. Sonneveld [180] observed that we can also construct the vectors $\widehat{r}_j = P_j^2(A) r_0$, using only the latter form of the inner product for recovering the bi-conjugate gradients parameters (which implicitly define the polynomial P_j). By doing so, the computation of the vectors $\widetilde{r}_j$ can be avoided and so can the multiplication by the matrix A^T.

We will make this more precise. In Bi-CG (Figure 7.1) we concentrated on the recursions that are necessary to compute the new residual vector r_i (for simplicity we consider the unpreconditioned algorithm: $K = I$):

$$p_i = r_{i-1} + \beta_{i-1} p_{i-1}$$
$$r_i = r_{i-1} - \alpha_i A p_i.$$

Now we express these vectors explicitly as members of the Krylov subspace, by using the polynomial expressions

$$r_i = P_i(A) r_0 \quad \text{and} \quad p_i = Q_{i-1}(A) r_0. \tag{7.3}$$

In these expressions the index denotes the degree of the polynomials involved.

Exercise 7.3. *Show that the vectors r_i and p_i can be expressed as in (7.3).*

The recursion for r_i leads to the following recursion for the polynomials:

$$P_i(A) = P_{i-1}(A) - \alpha_i A Q_{i-1}(A), \tag{7.4}$$

and from the recursion for p_i we obtain

$$Q_{i-1}(A) = P_{i-1}(A) + \beta_{i-1} Q_{i-2}(A). \tag{7.5}$$

We are interested in the computation of the vector $\widehat{r}_i = P_i(A)^2 r_0$, and therefore we square the expression in (7.4):

$$P_i(A)^2 = P_{i-1}(A)^2 + \alpha_i^2 A^2 Q_{i-1}(A)^2 - 2\alpha_i A P_{i-1}(A) Q_{i-1}(A). \tag{7.6}$$

In order to make this a computable recursion, we also need expressions for $Q_{i-1}(A)^2$ and for $P_{i-1}(A) Q_{i-1}(A)$. Squaring (7.5) leads to

$$Q_{i-1}(A)^2 = P_{i-1}(A)^2 + \beta_{i-1}^2 Q_{i-2}(A)^2 + 2\beta_{i-1} P_{i-1}(A) Q_{i-2}(A). \tag{7.7}$$

The expression for $P_{i-1}(A) Q_{i-1}(A)$ is obtained by multiplying (7.5) with $P_{i-1}(A)$:

$$P_{i-1}(A) Q_{i-1}(A) = P_{i-1}(A)^2 + \beta_{i-1} P_{i-1}(A) Q_{i-2}(A). \tag{7.8}$$

We need (7.7) and (7.8) for $P_{i-1}(A) Q_{i-2}(A)$, and this is obtained from the multiplication of the expression in (7.4) by $Q_{i-2}(A)$:

$$P_{i-1}(A) Q_{i-2}(A) = P_{i-2}(A) Q_{i-2}(A) - \alpha_{i-1} A Q_{i-2}(A)^2. \tag{7.9}$$

In the actual computations, all of these polynomial expressions have to operate on r_0. This leads to a set of vector recursions for the vectors

$$\begin{aligned} r_i &\equiv P_i(A)^2 r_0 \\ p_i &\equiv Q_{i-1}(A)^2 r_0 \\ u_i &\equiv P_{i-1}(A)Q_{i-1}(A) \\ q_i &\equiv P_i(A)Q_{i-1}(A). \end{aligned}$$

Exercise 7.4. *Show that the above relations lead to the algorithm given in Figure 7.3.*

x_0 is an initial guess, $r_0 = b - Ax_0$
$\widetilde{r}$ is an arbitrary vector, such that
$\widetilde{r}^T r_0 \neq 0$,
e.g., $\widetilde{r} = r_0$
for $i = 1, 2, \ldots.$
  $\rho_{i-1} = \widetilde{r}^T r_{i-1}$
  if $\rho_{i-1} = 0$ method fails
  if $i = 1$
    $u_1 = r_0$
    $p_1 = u_1$
  else
    $\beta_{i-1} = \rho_{i-1}/\rho_{i-2}$
    $u_i = r_{i-1} + \beta_{i-1} q_{i-1}$
    $p_i = u_i + \beta_{i-1}(q_{i-1} + \beta_{i-1} p_{i-1})$
  endif
  solve $\widehat{p}$ **from** $K\widehat{p} = p_i$
  $\widehat{v} = A\widehat{p}$
  $\alpha_i = \frac{\rho_{i-1}}{\widetilde{r}^T \widehat{v}}$
  $q_i = u_i - \alpha_i \widehat{v}$
  solve $\widehat{u}$ **from** $K\widehat{u} = u_i + q_i$
  $x_i = x_{i-1} + \alpha_i \widehat{u}$
  $r_i = r_{i-1} - \alpha_i A\widehat{u}$
  check convergence; continue if necessary
end

Figure 7.3. CGS algorithm.

In Bi-CG, the coefficient β_{i-1} is computed from the ratio of two successive values of ρ. We now have that

$$\begin{aligned}\rho_{i-1} &= \widetilde{r}_{i-1}^T r_{i-1}\\ &= (P_{i-1}(A^T)\widetilde{r}_0)^T P_{i-1}(A) r_0\\ &= \widetilde{r}_0^T P_{i-1}(A)^2 r_0\\ &= \widetilde{r}_0^T r_{i-1},\end{aligned} \tag{7.10}$$

which shows that the Bi-CG recursion coefficient can be recovered without forming the shadow residuals $\widetilde{r}_j$ as in Bi-CG.

Exercise 7.5. *Show that α_i can also be computed from the above defined vectors, that is, without forming the Bi-CG shadow residuals.*

In the CGS algorithm we have avoided working with A^T, which may be an advantage in some situations. However, the main attraction is that we are able to generate residuals for which

$$r_i = P_i(A)^2 r_0.$$

Note that it is easy to find the update formula for x_i.

Exercise 7.6. *Compare the computational costs for one iteration of Bi-CG with the costs for one iteration of CGS.*

In order to see the possible advantage of CGS, we have to realize that the operator $P_i(A)$ transforms the starting residual r_0 into a much smaller residual $r_i = P_i(A)r_0$, if Bi-CG converges well. We might hope that applying the operator once again gives another reduction in the residual. Indeed, the resulting CGS [180] method generally works very well for many unsymmetric linear problems. It often converges much faster than BI-CG (about twice as fast in some cases, because of the squaring effect of the operator $P_i(A)$). The surprising observation is that we obtain this faster convergence for about the same amount of work per iteration as in Bi-CG. Bi-CG and CGS have the advantage that fewer vectors are stored than in GMRES. These three methods have been compared in many studies (see, e.g., [160, 34, 158, 146]).

CGS, however, usually shows a very irregular convergence behaviour. This behaviour can even lead to cancellation and a 'spoiled' solution [201]; see also Chapter 8. Freund [87] suggested a squared variant of QMR, which was called TFQMR. His experiments showed that TFQMR is not necessarily faster than CGS, but it certainly has a much smoother convergence behaviour.

The scheme in Figure 7.3 represents the CGS process for the solution of $Ax = b$, with a given preconditioner K.

In exact arithmetic, the α_j and β_j are the same constants as those generated by Bi-CG. Therefore, they can be used to compute the Petrov–Galerkin approximations for eigenvalues of A.

Exercise 7.7. *Construct real unsymmetric systems with known eigenvalues. Test CGS and Bi-CG for problems where all eigenvalues have positive real part (and zero or small imaginary parts). Compare the convergence plots. Do the same for spectra with larger imaginary parts and for indefinite problems.*

Exercise 7.8. *Compare CGS and GMRES for some test problems. CGS may be faster than GMRES in terms of iterations, but can it be faster in terms of matrix vector products? Can CGS be faster in terms of CPU-time? In what sorts of situation?*

7.4.1 Numerical illustrations

We have seen that CGS generates residuals r_j that can be expressed as $r_j = p_j(A)^2 r_0$, where $p_j(A)$ describes the contraction for Bi-CG. We see this effect most easily for situations where the eigenvalues are distributed uniformly over some interval. As an example, we take a diagonal matrix A of order 200 with diagonal elements uniformly distributed over $[1, 1000]$. We take the right-hand side b so that the solution x of $Ax = b$ has elements that all are 1 (which makes it easy to check the solution). In Figure 7.4 we have plotted the residuals for Bi-CG and CGS, and we see the squaring effect.

Of course, the squaring effect of CGS may also have unwanted effects, that is, it magnifies local peaks in the convergence history of Bi-CG. This is what we see when we repeat the example for $A_1 = A - 50I$. The matrix A_1 is now indefinite and that leads to slow convergence of Krylov subspace methods (remember that one wants the iteration polynomial p_j to be small in all eigenvalues, but at the same time $p_j(0) = 1$). We see the convergence history in Figure 7.5.

Indeed, we see the sharper, and relatively higher, peaks for CGS in comparison with Bi-CG, and in fact we do not see much advantage for CGS in this case. However, if we inspect the process in more detail, then there may be advantages. As an example, we again consider the matrix A_1, but now we replace the largest diagonal element by 1050. Because the first $n - 1$ eigenvalues are now in the interval $[-49, 950]$, the n-th one is relatively well separated. Hence, we may expect that the eigenvector component of the residual corresponding

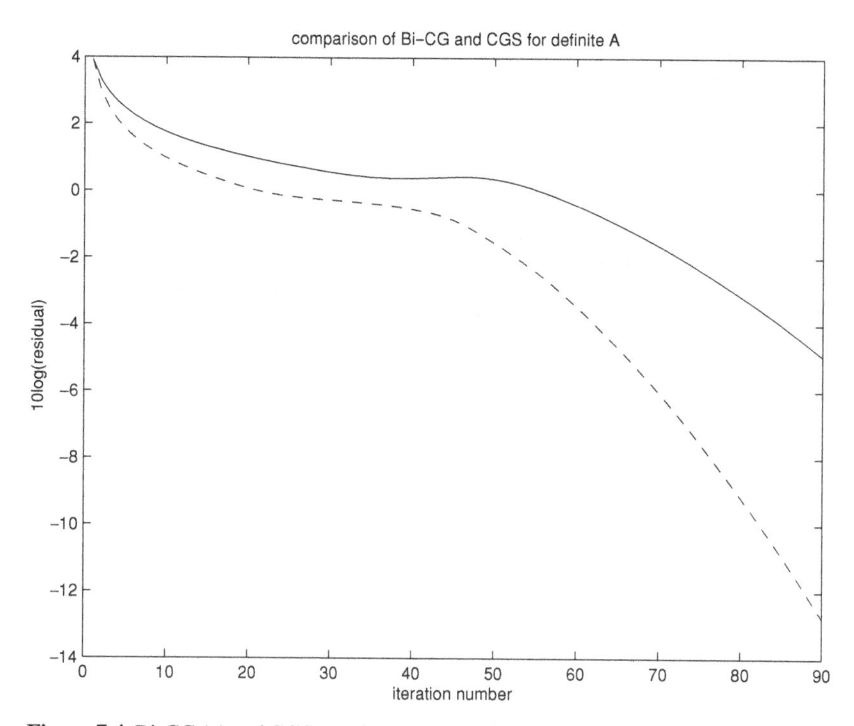

Figure 7.4. Bi-CG (–) and CGS (- -) for a system with uniformly distributed eigenvalues.

to this isolated eigenvalue is discovered relatively early in the iteration process by Lanczos-based methods. In Figure 7.6 we have plotted the absolute values of the n-th component of the residual vectors for Bi-CG and CGS. Indeed, we see that CGS returns a residual with increased accuracy for the last component (and hence a much better error in the approximated solution in this case).

In Figure 7.7 we have plotted the history for the 5-th component of the residual and now we see that CGS gives no advantage over Bi-CG (as was to be expected from the convergence history displayed in Figure 7.5.

We believe that the squaring effect in components of the residual is responsible for the observation that CGS often does remarkably well as an iterative process for the computation of the correction in a Newton iteration process for nonlinear systems (cf. [84, p.125].

7.5 Complex symmetric systems

In some applications the system matrix A is complex symmetric and not Hermitian. This occurs, for instance, in the modelling of electric currents in

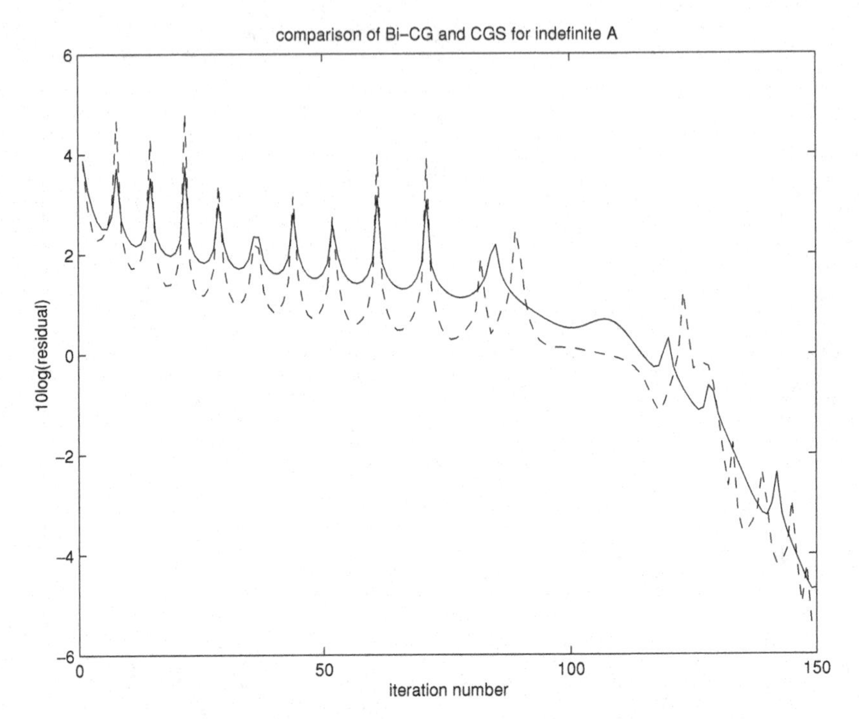

Figure 7.5. Bi-CG (–) and CGS (- -) for an indefinite system.

magnetic fields. The symmetry cannot be exploited in the standard way, that is with CG, MINRES, or SYMMLQ, in order to obtain short-term recurrence relations. However, three-term recurrence relations for a basis for the Krylov subspace can be derived by replacing the inner product for complex vectors v and w:

$$v^H w = \sum_{i=1}^{n} \bar{v}_i w_i,$$

by the bilinear form:

$$v^T w = \sum_{i=1}^{n} v_i w_i. \tag{7.11}$$

Exercise 7.9. *Show that (7.11) does not define a proper inner product over $\mathbb{C}^n$. In particular, determine a vector $v \neq 0$ for which $v^T v = 0$.*

Note that (7.11) represents the standard complex inner product for vectors $\bar{v}$ and w. This can be used to show that a complex symmetric matrix ($A = A^T$)

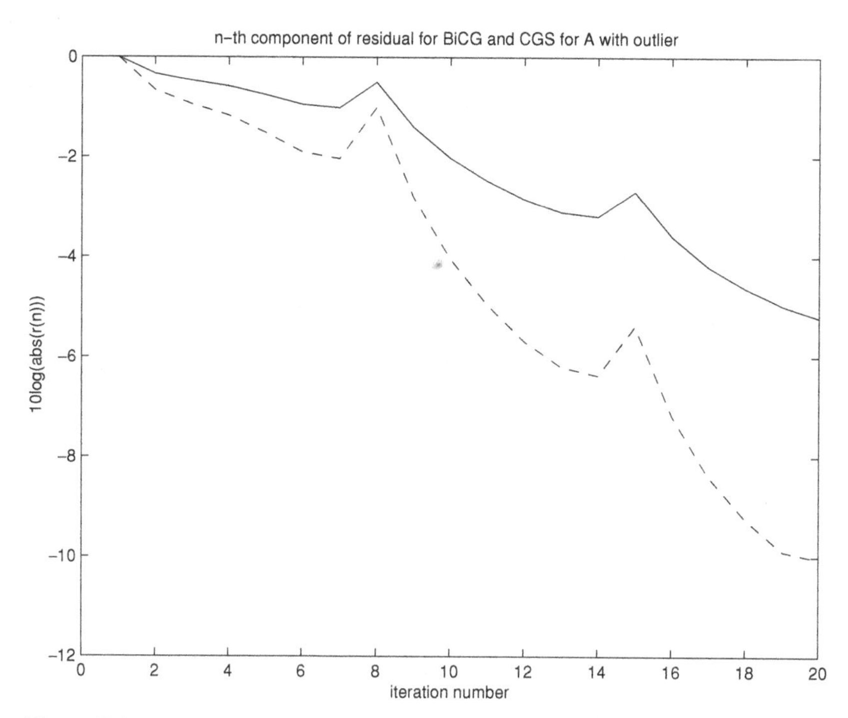

Figure 7.6. The n-th residual component Bi-CG (–) and CGS (- -) for an indefinite system with a well-separated eigenvalue.

is symmetric with respect to the bilinear form:

$$v^T Aw = \bar{v}^H Aw = \bar{A}^T \bar{v}^H w = (\bar{A}\bar{v})^H w = (Av)^T w.$$

We can now generate a basis for the Krylov subspace with a three term recurrence relation, just as we have done for the Conjugate Gradient method, by using the bilinear form (7.11) [204, 85]. If this process does not break down within m steps, then it generates a basis $v_1, v_2, \ldots, v_m$ that is not orthogonal, but satisfies the *conjugate orthogonality* relations

$$\bar{v}_i^H v_j = 0 \quad \text{for } i \neq j.$$

If we collect these vectors v_j in a matrix V_m, then we have the situation that V_m is, with the standard complex inner product, orthogonal with respect to $\bar{V}_m$. For this reason, the resulting method belongs to the class of Petrov–Galerkin methods, and in [204] the method has been named the *Conjugate Orthogonal Conjugate Gradient* method: COCG, for short.

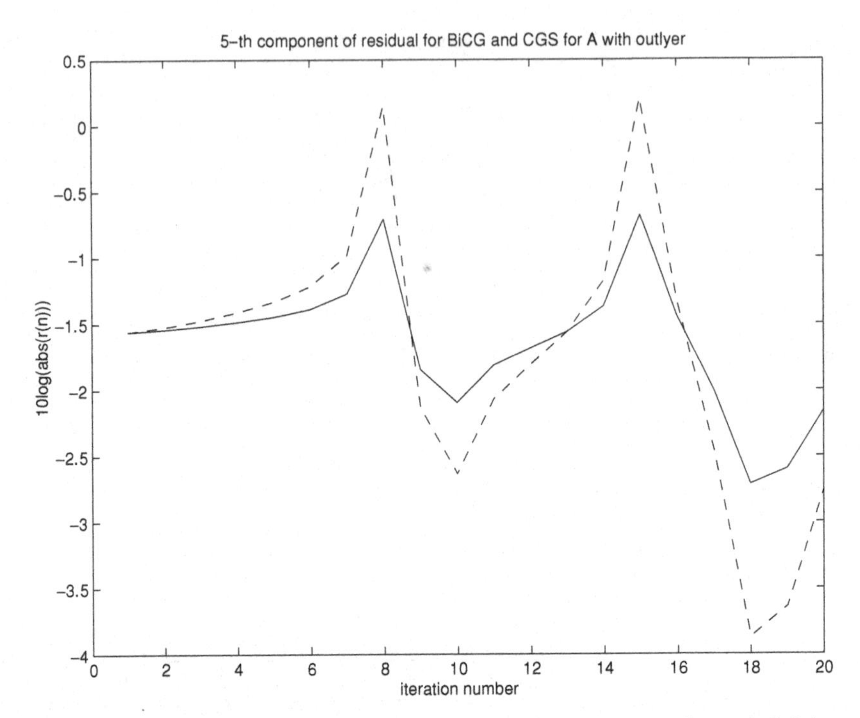

Figure 7.7. The 5-th residual component Bi-CG (–) and CGS (- -) for an indefinite system with a well-separated eigenvalue.

The method can be described by the algorithm given in Figure 7.8. In that algorithm we have included a preconditioner K, that also has to be (complex) symmetric.

Note that the only essential change with respect to the Conjugate Gradient method (cf. Figure 5.2) is the replacement of the standard complex inner product by the bilinear form (7.11). In this scheme for COCG we have assumed that no pivoting was necessary for the implicitly generated tridiagonal matrix T_m. Singularity of T_m can be tested by checking whether an intermediate $p_i^T q_i = 0$. Likewise, when this quantity is very small with respect to $\|p_i\|_2 \|Aq_i\|_2$, we may expect stability problems. Of course, we may also encounter a breakdown condition as that in Bi-CG, since we cannot be certain that $r_i^T r_i$ is very small with respect to $\|r_i\|_2^2$.

The pivoting problem could be circumvented by a QR decomposition of T, as in MINRES or SYMMLQ [85]. The other breakdown condition could be circumvented by a look-ahead strategy as has been suggested for Bi-CG. An easier, although possibly less efficient, fix would be to restart or to proceed with

```
x_0 is an initial guess, r_0 = b - A x_0
for i = 1, 2, ....
    Solve K w_{i-1} = r_{i-1}
    ρ_{i-1} = r_{i-1}^T w_{i-1}
    if i = 1
        p_i = w_{i-1}
    else
        β_{i-1} = ρ_{i-1} / ρ_{i-2}
        p_i = w_{i-1} + β_{i-1} p_{i-1}
    endif
    q_i = A p_i
    α_i = ρ_{i-1} / p_i^T q_i
    x_i = x_{i-1} + α_i p_i
    r_i = r_{i-1} - α_i q_i
    if x_i accurate enough then quit
end
```

Figure 7.8. COCG with preconditioning K.

another Krylov subspace method, thereby giving up the possible savings due to the complex symmetry.

Freund [85] has proposed using the QR decomposition for T_m, similar to MINRES and SYMMLQ, and solving the reduced system with the quasi-minimum residual approach as in QMR, see Section 7.3. His variant is based on Lanczos and generates, after k iterations, a symmetric tridiagonal matrix

$$T_{k+1,k} = \begin{bmatrix} \alpha_1 & \beta_2 & 0 & \cdots & 0 \\ \beta_2 & \alpha_2 & \beta_3 & \ddots & \vdots \\ 0 & \beta_3 & \ddots & \ddots & 0 \\ \vdots & \ddots & \ddots & \ddots & \beta_k \\ 0 & \cdots & 0 & \beta_k & \alpha_k \\ 0 & \cdots & \cdots & 0 & \beta_{k+1} \end{bmatrix}.$$

This matrix is rotated by successive Givens rotations

$$\begin{bmatrix} c_j & \bar{s}_j \\ -s_j & c_j \end{bmatrix}$$

$x_0 \in \mathbb{C}^n$ is an initial guess, $\widetilde{v} = b - Ax_0$
$v_0 = p_0 = p_{-1} = 0$
$\beta_1 = (\widetilde{v}^T\widetilde{v})^{1/2}$, $\bar{\tau}_1 = \beta_1$, $c_0 = c_{-1} = 1$,
$s_0 = s_{-1} = 0$
for $i = 1, 2, \ldots.$
  if $\beta_i = 0$ STOP: x_{i-1} solves $Ax = b$
  else $v_i = \widetilde{v}/\beta_i$, $\alpha_i = v_i^T A v_i$
  endif
  $\widetilde{v} = Av_i - \alpha_i v_i - \beta_i v_{i-1}$, $\beta_{i+1} = (\widetilde{v}^T\widetilde{v})^{1/2}$
  $\theta_i = \bar{s}_{i-2}\beta_i$, $\eta_i = c_{i-1}c_{i-2}\beta_i + \bar{s}_{i-1}\alpha_i$
  $\mu = c_{i-1}\alpha_i - s_{i-1}c_{i-2}\beta_i$, $|\zeta_i| = (|\mu|^2 + |\beta_{i+1}|^2)^{1/2}$
  if $\mu = 0$
    $\zeta_i = |\zeta_i|$
  else $\zeta_i = |\zeta_i|\mu/|\mu|$
  endif
  $c_i = \mu/\zeta_i$, $s_i = \beta_{i+1}/\zeta_i$
  $p_i = (v_i - \eta_i p_{i-1} - \theta_i p_{i-2})/\zeta_i$
  $\tau_i = c_i\bar{\tau}_i$, $\bar{\tau}_{i+1} = -s_i\bar{\tau}_i$
  $x_i = x_{i-1} + \tau_i p_i$
  if x_i is accurate enough **then** quit
end

Figure 7.9. QMR-SYM for complex symmetric systems.

to upper triangular form $R_{k+1,k}$:

$$R_{k+1,k} = \begin{bmatrix} \zeta_1 & \eta_2 & \theta_3 & 0 & \cdots & 0 \\ 0 & \zeta_2 & \eta_3 & \ddots & \ddots & \vdots \\ 0 & \ddots & \zeta_3 & \ddots & \ddots & 0 \\ \vdots & & \ddots & \ddots & \ddots & \theta_k \\ \vdots & & & \ddots & \ddots & \eta_k \\ 0 & \cdots & \cdots & \cdots & 0 & \zeta_k \\ 0 & \cdots & \cdots & \cdots & \cdots & 0 \end{bmatrix}.$$

We recognize all these elements in the algorithm QMR-SYM [85], given in Figure 7.9. Although this algorithm is more complicated than COCG, given before, we have included it, because it avoids one of the possible breakdowns of COCG. This algorithm might be made even more robust by adding look-ahead strategies, but it may be as efficient to restart at the iteration just before breakdown, because this event is rather rare. Note that we have given the algorithm with weights $\omega_j = 1$. For a weighted minimization variant see [85, Algorithm 3.2].

We have seen that these variants of Bi-CG for complex symmetric systems can be derived from either Conjugate Gradients or the Lanczos procedure by using the bilinear form $v^T w$ instead of the proper inner product $v^H w$ over complex vector spaces. The question may arise – how safe is it to work with a 'wrong' inner product? Freund gives an elegant argument for the usefulness of QMR-SYM [85, p.428]: any complex n by n matrix is similar to a complex symmetric matrix. This result implies that the general nonsymmetric Lanczos method differs from the complex symmetric one only in an additional starting vector $\widetilde{r}_0$, which can be chosen independently from r_0. In [85] numerical experiments are reported in which QMR-SYM compares favourably with GMRES(m), Bi-CG, and CGS.

8

How serious is irregular convergence?

Bi-CG and methods derived from Bi-CG can display rather irregular convergence behaviour. By irregular convergence I refer to the situation where successive residual vectors in the iterative process differ in orders of magnitude in norm, and some of these residuals may even be much bigger in norm than the starting residual. In particular the CGS method suffers from this phenomenon. I will show why this is a point of concern, even if eventually the (updated) residual satisfies a given tolerance.

In the Bi-CG algorithms, as well as in CG, we typically see in the algorithm a statement for the update of x_i, such as

$$x_{i+1} = x_i + w_i \tag{8.1}$$

and a statement for the update of r_i, of the form

$$r_{i+1} = r_i - Aw_i. \tag{8.2}$$

We see that, in exact arithmetic, the relation $r_{i+1} = b - Ax_{i+1}$ holds, just as expected. A further inspection of these algorithms reveals that x_i is not used at other places in the basic algorithm, whereas the r_i is also used for the computation of the search direction and for iteration parameters. The important consequence of this is that rounding errors introduced by the actual evaluation of r_{i+1} using equation (8.2) will influence the further iteration process, but rounding errors in the evaluation of x_{i+1} by (8.1) will have no effect on the iteration. This would not be much of a problem if the rounding error

$$\delta r_{i+1} \equiv fl(r_i - Aw_i) - (r_i - Aw_i)$$

would match the rounding error

$$\delta x_{i+1} \equiv fl(x_i + w_i) - (x_i + w_i),$$

in the sense that $\delta r_{i+1} = -A\delta x_{i+1}$, since that would keep the desired relation $r_{i+1} = b - Ax_{i+1}$ intact. However, it will be obvious that this is unrealistic, and the question remains – how serious can a possible deviation between r_j and $b - Ax_j$ be?

Of course, we make rounding errors in (8.1) and (8.2) through the vector addition, but usually these errors will be small in comparison with the rounding errors introduced in the multiplication of w_i with A. Therefore, we will here consider only the effect of these errors. In this case, we can write the computed r_{i+1} as

$$r_{i+1} = r_j - Aw_i - \Delta_{A_i} w_i, \tag{8.3}$$

where Δ_A is an $n \times n$ matrix for which $|\Delta_{A_i}| \preceq n_A \overline{\xi}\, |A|$: n_A is the maximum number of nonzero matrix entries per row of A, $|B| \equiv (|b_{ij}|)$ if $B = (b_{ij})$, $\overline{\xi}$ is the relative machine precision, the inequality $\preceq$ refers to element-wise $\leq$.

It then simply follows that

$$r_k - (b - Ax_k) = \sum_{j=1}^{k} \Delta_{A_j} w_j = \sum_{j=1}^{k} \Delta_{A_j}(e_{j-1} - e_j), \tag{8.4}$$

e_j is the approximation error in the j-th approximation: $e_j \equiv x - x_j$. Hence,

$$\begin{aligned} \big| \|r_k\| - \|b - Ax_k\| \big| &\leq 2k\, n_A \overline{\xi}\; \||A|\| \max_j \|e_j\| \\ &\leq 2k\, n_A \overline{\xi}\; \||A|\|\; \left\|A^{-1}\right\| \max_j \|r_j\|. \end{aligned} \tag{8.5}$$

Except for the factor k, the first upperbound appears to be rather sharp. We see that an approximation with a large approximation error (and hence a large residual) may lead to inaccurate results in the remaining iteration process. Such large local approximation errors are typical for CGS, and van der Vorst [201] describes an example of the resulting numerical inaccuracy. If there are a number of approximations with comparable, large approximation errors, then their multiplicity may replace the factor k, otherwise it will be only the largest approximation error that makes up virtually all of the bound for the deviation.

For more details we refer to Sleijpen and van der Vorst [176], Sleijpen et al. [177].

Exercise 8.1. *Show that similar update expressions can also be formulated for Bi-CGSTAB. Is the difference essential for the above discussion on the influence of rounding errors? Derive an upperbound for the deviation between* $\|r_k\|_2$ *and* $\|b - Ax_k\|_2$ *for Bi-CGSTAB, again under the assumption that only the multiplication with A leads to rounding errors.*

8.1 Reliable updating

It is of course important to maintain a reasonable correspondence between r_k and $b - Ax_k$, and the easiest way to do this would be to replace the vector r_k by $b - Ax_k$. However, the vectors r_k steer the entire iterative process and their relation defines the projected matrix $T_{i,i}$. If we replace these vectors then we ignore the rounding errors to these vectors and it will be clear that the iteration process cannot compensate for these rounding errors. They may be significant at iteration steps where the update to r_j is relatively large and the above sketched naive replacement strategy may not then be expected to work well. Indeed, if we replace r_{i+1} in CGS by $b - Ax_i$, instead of updating it from r_i, then we observe stagnation in the convergence in many important situations. This means that we have to be more careful.

Neumaier (see references in [176]) suggested replacing r_j by $b - Ax_j$ in CGS only at places where $\|r_j\|_2$ is smaller than the smallest norm of the residual in the previous iteration history and carrying out a groupwise update for the iterates in between. Schematically, the groupwise update and residual replacement strategy of Neumaier can be described as in Figure 8.1.

This scheme was further analysed and refined, in particular with a flying restart strategy, in [176]. Note that the errors in the evaluation of w_j itself are not so important: it is the different treatment of w_j in the updating of x_j and of r_j that causes the two vectors to lose the desired mutual relation. In this respect we may consider the vectors w_j as exact quantities.

Groupwise Solution Update:
$z = x_0$, $\widehat{x} = 0$, $r_{min} = \|r_0\|_2$
...
for $j = 0, 2, \ldots,$ until convergence
    ...
    $\widehat{x} = \widehat{x} + w_j$ (instead of update of x_i)
    if $\|r_j\|_2 < r_{min}$ (i.e. group update)
        $z = z + \widehat{x}$
        $\widehat{x} = 0$
        $r_j = b - Az$
        $r_{min} = \|r_j\|_2$
    end if
end

Figure 8.1. Neumaier's update strategy.

At a replacement step we perturb the recurrence relation for the basis vectors of the Krylov subspace and we want these errors to be as small as possible. The updates w_j usually vary widely in norm in various stages of the iteration process, for instance in an early phase these norms may be larger than $\|r_0\|_2$, whereas they are small in the final phase of the iteration process. Especially in a phase between two successive smallest values of $\|r_j\|_2$, the norms of the updates may be a good deal larger than in the next interval between two smallest residual norms. Grouping the updates avoids rounding errors within one group spoiling the result for another group. More specifically, if we denote the sum of w_js for the groups by S_i, and the total sum of updates by S, then groupwise updating leads to errors of the magnitude of $\xi|S_i|$, which can be much smaller than $\xi|S|$.

Now we have to determine how much we can perturb the recurrence relations for the Lanczos vectors r_j. This has been studied in much detail in [189]. It has been observed by many authors that the driving recurrences $r_j = r_{j-1} - \alpha_{j-1}Aq_{j-1}$ and $q_j = r_j + \beta_{j-1}q_{j-1}$ are locally satisfied almost to machine precision and this is one of the main properties behind the convergence of the computed residuals [100, 189, 176]. Tong and Ye [189] observed that these convergences are maintained even when we perturb the recurrence relations with perturbations that are significantly greater than machine precision, say of the order of the square root of the machine precision ξ, in a relative sense.

The idea, presented in [208], is to compute an upperbound for the deviation in r_j, with respect to $b - Ax_j$, in finite precision, and to replace r_j by $b - Ax_j$ as soon as this upperbound reaches the relative level of $\sqrt{\xi}$. This upperbound is denoted by d_j and it is computed from the recurrence

$$d_j = d_{j-1} + \xi N \|A\| \|\widehat{x}_j\| + \xi \|r_j\|,$$

with N the maximal number of nonzero entries per row of A.

The replacement strategy for reliable updating is then implemented schematically as in Figure 8.2.

Remark: For this reliable implementation, we need to put a value for N (the maximal number of nonzero entries per row of A) and $\|A\|$. The number of nonzero entries may, in applications, vary from row to row, and selecting the maximum number may not be very realistic. In my experience with sparse matrices, the simple choice $N = 1$ still leads to a practical estimate d_n for $\|\delta_n\|$. For $\|A\|$, I suggest simply taking $\|A\|_\infty$.

In any case, we note that precise values are not essential, because the replacement threshold ϵ can be adjusted. We also need to choose this ϵ. Extensive numerical testing (see [208]) suggests that $\epsilon \sim \sqrt{\xi}$ is a practical criterion. However, there are examples where this choice leads to stagnating residuals at some

Reliable Updating Strategy:
Input: an initial approximation $x = x_0$
a residual replacement threshold $\epsilon = \sqrt{xi}$, an estimate of $N\|A\|$
$r_0 = b - Ax_0$ $\widehat{x}_0 = 0$, $d_{init} = d_0 = \xi(\|r_0\| + N\|A\|\|x_0\|)$,
for $j = 1, 2, \ldots,$ until convergence
Generate a correction vector q_j by the Iterative Method
$\widehat{x}_j = \widehat{x}_{j-1} + q_j$
$r_j = r_{j-1} - Aq_j$
$d_j = d_{j-1} + \xi N\|A\|\ \|\widehat{x}_j\| + \xi\ \|r_j\|$
if $d_{j-1} \leq \epsilon\|r_{j-1}\|,\ d_j > \epsilon\|r_j\|$ and $d_j > 1.1 d_{init}$
$z = z + \widehat{x}_j$
$\widehat{x}_j = 0$
$r_j = b - Az$
$d_{init} = d_j = \xi(\|r_j\| + N\|A\|\|z\|)$
end if
end for
$z = z + \widehat{x}_j$

Figure 8.2. Ye–van der Vorst update strategy.

unacceptable level. In such cases, choosing a smaller ϵ will regain the convergence to $O(\xi)$.

The presented implementation requires one extra matrix-vector multiplication when a replacement is carried out. Since only a few steps with replacement are required, this extra cost is marginal relative to the other costs. However, some savings can be made by selecting a slightly smaller ϵ and carrying out residual replacement at the step next to the one for which the residual replacement criterion is satisfied (cf. [176]). It also requires one extra vector storage for the groupwise solution update (for z) and computation of a vector norm $\|\widehat{x}_n\|$ for the update of d_n ($\|r_n\|$ is usually computed in the algorithm for stopping criteria).

8.2 Rounding errors and discretization errors

When a linear system $Ax = b$ is solved in rounded finite precison floating point arithmetic, then the obtained solution $\widehat{x}$ is most likely contaminated by rounding errors. In the absence of precise information about these errors, we have to work with bounds for their effects. A well-known rigorous bound for the relative error in the solution of the perturbed linear system

$$(A + \delta A)y = b + \delta b$$

is given by [98, Chapter 2.7.3]:

$$\frac{\|y - x\|}{\|x\|} \leq \frac{2\epsilon}{1 - r}\kappa(A), \tag{8.6}$$

under the conditions that

$$\|\delta A\| \leq \epsilon \|A\|, \quad \|\delta b\| \leq \epsilon \|b\|, \quad \epsilon\kappa(A) = r < 1,$$

and $\kappa(A)$ denotes the condition number of A:

$$\kappa(A) \equiv \|A^{-1}\| \, \|A\|.$$

The upperbound is quite realistic in the sense that there exist perturbations that lead to errors that are quite close to the upperbound.

However, this does not imply that we may expect the multiplying effect of the condition number for perturbations that have a specific structure, such as discretization errors. In particular, if $Ax = b$ comes from a standard finite difference or finite element discretization of an elliptic partial differential equation, then a common situation is that the exact solution $\bar{x}$ of the PDE, restricted to the gridpoints, satisfies

$$A\bar{x} = b + \tau,$$

with $\tau_i = \mathcal{O}(h^2)$.

In such situations, the solution x of $Ax = b$ typically satisfies

$$\|x - \bar{x}\| = \mathcal{O}(h^2).$$

See, for instance, [112, Chapter 4.5] or [159, Chapter 6.2.1] for more detailed information on this.

This means that a structured error in b may have a very well-understood effect on the solution, in contrast to unstructured errors. When the discretized system $Ax = b$ is represented by machine numbers, then we introduce an additional relative error in the solution that is bounded by (8.6), with ϵ replaced by the relative machine precision ξ.

A major source of errors in the actually computed solution is in the algorithm by which the solution is generated. Gaussian elimination may lead to large additional errors, even with partial pivoting [98, Chapter 3.3.2]. The errors introduced by the elimination process can be largely removed by iterative refinement, under the condition that the residual $b - A\widehat{x}$ can be computed with some significant digits [98, Chapter 3.5.3]. It is tempting to believe that this is automatically the case in iterative solution methods, but this is only the case if the residual is computed explicitly, at least from time to time, as in the reliable updating technique. In particular, for an explicitly computed residual $r_i = b - Ax_i$, we have with (8.6) that

$$\frac{\|x_i - x\|}{\|x\|} \leq \frac{2}{1-r}\kappa(A)\frac{\|r_i\|}{\|b\|},$$

under the condition that $\kappa(A)\frac{\|r_i\|}{\|b\|} = r < 1$.

8.3 Effects of rounding errors to Krylov processes

In this section we will study the possible effects of inexact computations on Krylov subspace methods. We have seen how the effects of large updates (or better still large residual vector corrections) can be reduced by appropriate update techniques. Now we will focus on other parts of the algorithms.

Essentially, the Krylov subspace methods are composed with the following ingredients

(a) The construction of a (bi-)orthogonal basis,
(b) The solution of a small reduced linear system,
(c) The construction of an approximate solution as a combination of the basis vectors.

Each of these three elements is a source for rounding errors and the fact that these elements are usually mixed in efficient implementations makes the analysis complicated. As a result, rigid upperbounds for actual methods are usually pessimistic and do not describe the actually observed behaviour well. For an example of state of the art work with respect to GMRES, see [64].

We will follow a slightly different approach in order to gain more insight into the main sources of rounding errors. We will restrict ourselves to symmetric matrices. In that case, a basis for the Krylov subspace is generated by the Lanczos method. This remarkable three-term recurrence relation does many things simultaneously: it provides a reduced system that can be used to generate an approximate solution and it contains useful spectral information on the matrix A. Even more remarkable is its behaviour with respect to rounding errors. In finite arithmetic the iteration coefficients may differ by more than 100% from those in exact arithmetic, after a number of iteration steps. Nevertheless, the reduced system may still lead to accurate approximate solutions and to accurate spectral information.

In the early 1950s, the Lanczos method was regarded as a method for the reduction of a square matrix A to tridiagonal form, and after the full n steps a similarity transformation of A to tridiagonal form was expected, if no early termination was encountered. In the latter case an invariant subspace would have been identified, which information is just as useful. It was soon recognized that rounding errors could change the process dramatically. Engeli et al. [78] published an example for which the desired spectral information was obtained only after many more than n iterations. A decade later, Ginsburg [94] made an algorithmic description of the Conjugate Gradients that was included in the well-known Handbook of Automatic Computation [223]. As an illustrative example, Ginsburg also used a discretized bi-harmonic equation for a one-dimensional beam problem, which led to the 40 by 40 matrix:

$$A = \begin{bmatrix} 5 & -4 & 1 & & & & & & & & \\ -4 & 6 & -4 & 1 & & & & & & & \\ 1 & -4 & 6 & -4 & 1 & & & \emptyset & & & \\ & 1 & -4 & 6 & -4 & 1 & & & & & \\ & & \cdot & \cdot & \cdot & \cdot & \cdot & & & & \\ & & & \cdot & \cdot & \cdot & \cdot & \cdot & & & \\ & & & & \cdot & \cdot & \cdot & \cdot & \cdot & & \\ & & & & & \cdot & \cdot & \cdot & \cdot & \cdot & \\ & & & & & 1 & -4 & 6 & -4 & 1 & \\ & \emptyset & & & & & 1 & -4 & 6 & -4 & 1 \\ & & & & & & & 1 & -4 & 6 & -4 \\ & & & & & & & & 1 & -4 & 5 \end{bmatrix}.$$

Then CG was used to solve the linear system $Ax = b$, where b is the first unit vector (of length 40). The convergence history is shown in Figure 8.3. Clearly, the CG method requires about 90 iterations to obtain a fairly accurate solution, and the approximate solution after 40 iterations is quite far from what it should have been in exact computation.

In the same figure, we have also plotted the convergence history for FOM, where we have solved the reduced system with Givens transformations in order to have optimal numerical stability. In theory, FOM and CG should produce the same iterates and we see that this is more or less the case until the 35-th iteration step. The main difference between FOM and CG is that in FOM we orthogonalize the new basis vector for the Krylov subspace against all previous

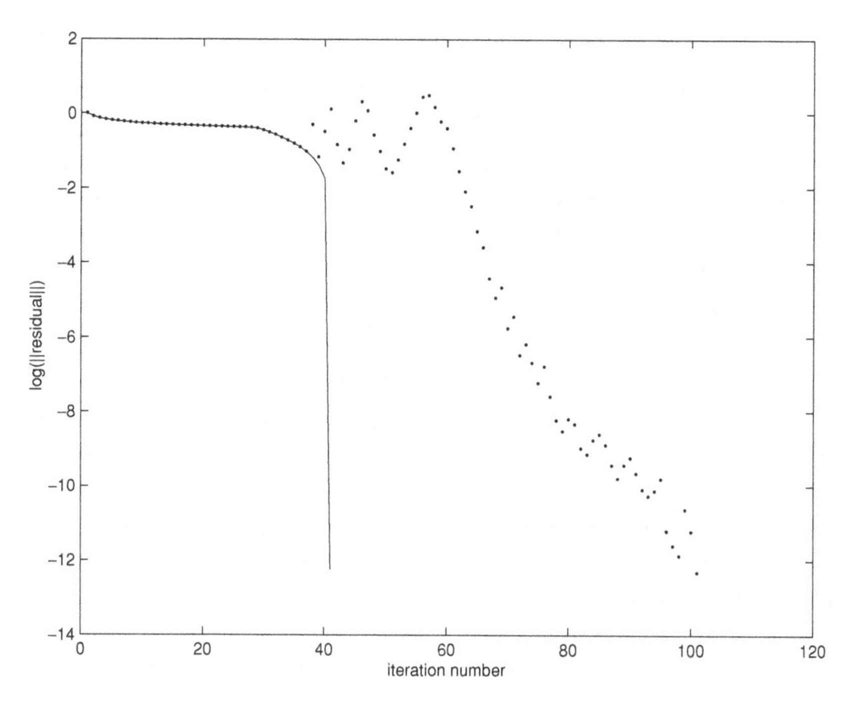

Figure 8.3. CG (...) and FOM (–).

basis vectors. Apparently, the loss of orthogonality, among the basis vectors generated with the 3-term recurrence relations, leads to a significant delay of the convergence. This was not understood in the early 1970s so that it is no surprise that CG was not very popular around that time. The CG subroutine, published in [223], was one of the very few that did not make it to the influential LINPACK and EISPACK software collections. It was only in the late 1990s that the conjugate gradients method entered Matlab.

8.3.1 The Lanczos recurrence in finite precision

In our attempts to understand the main effects of rounding errors we initially follow Greenbaum's analysis [101, Chapter 4]. This analysis starts with the famous results of Paige [150], published in 1976. Paige gave a rigorous analysis of the behaviour of the Lanczos algorithm in finite precision. Before we are in a position to quote an important theorem from [150], we have to introduce some notations.

Let ξ denote the floating point relative machine precision. With m we denote the maximum number of nonzero elements per row of the n by n matrix A.

We repeat the algebraic formulation, in exact arithmetic, of the Lanczos reduction

$$AV_k = V_{k+1}T_{k+1,k}, \tag{8.7}$$

where $T_{k+1,k}$ is a $k+1$ by k tridiagonal matrix and V_j denotes the n by j matrix with the orthonormal Lanczos vectors $v_1, v_2, ..., v_j$ as its columns. This relation can equivalently be expressed as

$$AV_k = V_kT_k + t_{k+1,k}v_{k+1}^T e_k^T,$$

where e_k denotes the k-th unit vector of length k.

We know that the Lanczos vectors are, in exact arithmetic, generated with Arnoldi's algorithm, given in Figure 3.1. The k by k leading part of the reduced matrix $H_{k+1,k}$ is symmetric, expressed by the fact that we have replace H by T in (8.7). When working in finite precision, we have to be more careful in how we ensure that T is tridiagonal. First, we avoid, of course, the computation of the elements outside the tridiagonal part. Second, we ensure that the matrix T is symmetric, by avoiding again computing the lower diagonal elements of T. In fact, Paige [150] proved his results for the implementation shown in Figure 8.4 of the algorithm in Figure 3.1.

Let v_1 be given with $v_1^T v_1 = 1$
$u = Av_1$
for $j = 1, 2, ..., k$
  $\alpha_j = v_j^T u$
  $w = u - \alpha v_j$
  $\beta_j = \|w\|_2$, **if** $\beta_j = 0$ STOP
  $v_{j+1} = w/\beta_j$
  $u = Av_{j+1} - \beta_j v_j$
end for

Figure 8.4. The Paige-style Lanczos algorithm.

Exercise 8.2. *Show that the algorithm in Figure 8.4 leads in exact arithmetic to the same vectors v_j as the algorithm in Figure 3.1. How do the α_j and β_j define the matrix $T_{k+1,k}$?*

This sets the stage for the following results.

Theorem 8.1. Paige [150]. *Let A be an n by n real symmetric matrix with at most m nonzero elements in any row. If the Lanczos algorithm described in Figure 8.4 is executed in floating point arithmetic with relative machine precision ξ, then α_j, β_j, v_j will be computed such that*

$$AV_k = V_k T_k + \beta_k v_{k+1} e_k^T + \delta V_k \tag{8.8}$$

$$\text{with } V_k \equiv [v_1, ..., v_k]$$

$$\delta_k \equiv [\delta v_1, ..., \delta v_k]$$

$$|v_j^T v_j - 1| \leq (n+4)\xi \tag{8.9}$$

$$\|\delta v_j\|_2 \leq \|A\|_2(7 + m\| \, |A| \, \|_2/\|A\|_2)\xi. \tag{8.10}$$

Note that from now on, in this chapter, V_k and T_k will denote the computed quantities. We will now follow Greenbaum's way of reasoning and study what will happen if we use the computed V_k and T_K in a straightforward manner. We will consider two different cases:

(A) *The Ritz–Galerkin approach:* In this case the approximated solution is defined as

$$x_k^R = V_k y_k^R \quad \text{with} \quad y_k^R = \|r_0\|_2 T_{k,k}^{-1} e_1, \tag{8.11}$$

and we will use this definition for the computation. In fact, this definition is, in exact arithmetic, equivalent to the CG method, but then the x_k is computed from different formulas.

(B) *The Minimum Residual approach:* Here the approximated solution is defined by

$$x_k^M = V_k y_k^M, \tag{8.12}$$

where y_k^M is defined by

$$\min_y \| \, \|r_0\|_2 e_1 - T_{k+1,k} y\|_2 \, (= \|b - Ax_k^M\|_2).$$

In exact arithmetic, these definitions lead to an x_k^M that is the same as the one computed by MINRES, but, again, in MINRES other formulas are used.

We will now assume that the defining formulas (8.11) and (8.12) hold exactly for the computed V_k and $T_{k+1,k}$.

Analysis for case (A) The residual for x_k^R is:

$$\begin{aligned} r_k^R &= r_0 - AV_k y_k^R \\ &= r_0 - (V_k T_{k,k} + \beta_k v_{k+1} e_k^T + \delta V_k) y_k^R \\ &= -\beta_k v_{k+1} e_k^T y_k^R - \delta V_k y_k^R && (8.13) \\ &= -\beta_k \|r_0\|_2 v_{k+1} e_k^T T_{k,k}^{-1} e_1 - \delta V_k y_k^R. && (8.14) \end{aligned}$$

Note that we have treated y_k^R differently in the last expression. The reason is that in the first term at the right-hand side of (8.14) it is only the last term of the solution that matters and by using $T_{k,k}$ this can be nicely exploited. We will deviate slightly from Greenbaum's analysis by a different interpretation of the formulas.

From (8.14) it follows that the norm of $\|r_k^R\|_2$ can be bounded as

$$\|r_k^R\|_2 \le |\beta_k| \, \|r_0\|_2 \|v_{k+1}\|_2 \, |e_k^T T_{k,k}^{-1} e_1| + \|\delta V_k\|_2 \, \|y_k^R\|_2. \tag{8.15}$$

Exercise 8.3. *Consider the Lanczos process for the system*

$$T_{k+1,k+1} y = e_1. \tag{8.16}$$

Show that j steps of the process, starting with $v_1 = e_1$, generate the Lanczos basis $v_i = e_i$, for $i = 1, \ldots, j \le k$ and that the reduced matrix $\widetilde{T}$ is given by

$$\widetilde{T}_{j+1,j} = Tj + 1, j.$$

Exercise 8.4. *Show, with the result of exercise 8.3, that $j \le k$ steps with CG for the system (8.16), with starting vector $y_0 = 0$, lead to an approximate solution for which the norm of the residual satisfies:*

$$\|r_j\|_2 = \beta_j \, |e_j^T T_{j,j}^{-1} e_1|. \tag{8.17}$$

This last exercise shows that the first term in the right-hand side of (8.15) can be interpreted as $\|r_0\|_2$ times the norm of a residual of a CG process. It remains to link this CG process with a CG process for A. In exact arithmetic, the Lanczos algorithm leads to a transformation of A to tridiagonal form (or to a partial reduction with respect to an invariant subspace):

$$AV_n = V_n T_{n,n}.$$

Paige [151] has shown that in floating point arithmetic a tridiagonal matrix is generated that has eigenvalues θ_j that are in the interval

$$[\min \lambda(A) - \tau_1, \max \lambda(A) + \tau_2], \tag{8.18}$$

where τ_1, τ_2 are mild functions of j times the machine precision ξ. We can now view the matrix $T_{k+1,k}$ as a section of a much larger matrix T that satisfies Paige's eigenvalue bounds. This means that the residuals $\beta_k |e_k^T T_{k,k}^{-1} e_1|$ (see equation (8.17)) can be bounded as in CG (5.25), which leads to an upperbound that converges with a factor

$$\frac{\sqrt{\widetilde{\kappa}} - 1}{\sqrt{\widetilde{\kappa}} + 1},$$

where $\widetilde{\kappa}$ is the condition number based on the spectrum in (8.18).

This means that the first term in the right-hand side of (8.15) can be bounded by a function that converges to zero in almost the same way as the usual upperbound for the CG residuals for A (note that $\|v_{k+1}\|_2$ can be bounded essentially by 1 apart from a very small correction in the order of ξ). Eventually, the second term will dominate and this second term can be bounded as (using (8.10) and (8.11))

$$\| \, \|\delta V_k\|_2 \, \|y_k^R\|_2 \|_2 \leq \sqrt{k} \|A\|_2 (7 + m \| \, |A| \, \|_2 / \|A\|_2) \xi \, \|r_0\|_2 \|T_{k,k}^{-1}\|_2. \quad (8.19)$$

According to Paige's results, $\|T_{k,k}^{-1}\|_2$ is equal, apart from a very small correction, to $\|A^{-1}\|_2$. This leads to an upperbound that eventually behaves like a function that is proportional to $\sqrt{k}\kappa\xi$, which is rather satisfactory. The final conclusion is that the errors in the Lanczos process do not seriously affect the converging properties of the Ritz–Galerkin approach.

Analysis for (B) For the minimum residual case, we can follow exactly the same arguments as for case (A). Now we have for the residual for x_k^M (again following Greenbaum's analysis [101, Chapter 4.3],:

$$\begin{aligned} r_k^M &= r_0 - A V_k y_k^M \\ &= r_0 - (V_{k+1} T_{k+1,k} + \delta V_k) y_k^M \\ &= V_{k+1} (\|r_0\|_2 - T_{k+1,k} y_k^M - \delta V_k y_k^M. \end{aligned} \quad (8.20)$$

From (8.20) it follows that the norm of $\|r_k^M\|_2$ can be bounded as

$$\|r_k^M\|_2 \leq \|V_{k+1}\|_2 \, \| \, \|r_0\|_2 e_1 - T_{k+1,k} y_k^M \|_2 + \|\delta V_k\|_2 \, \|y_k^M\|_2. \quad (8.21)$$

The first term at the right-hand side goes to zero on account of the relation between Ritz–Galerkin (in exact arithmetic: FOM) and Minimum Residual (in exact arithmetic: MINRES) residuals, see (6.3). This implies that eventually the second term in the error bound (8.21) dominates and this is the same term as for the Ritz–Galerkin approach.

From this we conclude that the loss of accuracy in the Lanczos process has no serious effects on the overall convergence of the Ritz–Galerkin and the Minimum Residual iterative approaches. By no serious effects we mean that the global upperbound, based on the condition number of A, is almost the same in exact and in finite precision floating point arithmetic. However, effects may be observed. We have seen that, depending on the spectral distribution of A, superlinear convergence may occur for, for instance, CG. This increased speed of convergence happens when extreme eigenvalues have been approximated sufficiently well in the underlying Lanczos process. It is well known that loss of orthogonality in the Lanczos process goes hand in hand with the occurrence of (almost) multiple eigenvalues of $T_{k,k}$ (see, for instance, [155, Chapter 13.6], [150]). These so-called multiplets have nothing to do with possibly multiple eigenvalues of A. Because of rounding errors, the eigenvalues of A, to which eigenvalues of $T_{k,k}$ have converged, have their effect again on future iterations and they may reduce (part of) the effects of the so-called superlinear convergence behaviour. This is visible in a delay of the convergence with respect to the convergence that would have been observed in exact arithmetic (or that might have been expected from gaps in the spectrum of A).

8.3.2 Effects of rounding errors on implementations

In the previous section we have discussed the effects of finite precision computations in the Lanczos process and we have seen that these effects are limited. In actual implementations there may be bigger effects because of a specific implementation. This has been analysed to some extent in [179]. We will follow this discussion for two possible implementations of the minimum residual approach for symmetric linear systems.

We start again from (8.12). We see that a small overdetermined linear system

$$T_{k+1,k}y_k^M = \|r_0\|_2 e_1$$

has to be solved, and a good and stable way of doing this is via a reduction of $T_{k+1,k}$ with Givens rotations to upper triangular form

$$T_{k+1,k} = Q_{k+1,k}R_k,$$

in which R_k is a k by k upper triangular band matrix with bandwidth 3 and $Q_{k+1,k}$ is a $k+1$ by k matrix with orthonormal columns (that is the product of the k Givens rotations).

This leads to the approximated solution

$$x_k^M = \|r_0\|_2\, V_k R_k^{-1} Q_{k+1,k}^T e_1. \tag{8.22}$$

With the given Lanczos basis, this approximated solution can be computed in a straightforward manner by first evaluating the vector

$$z_k = Q_{k+1,k}^T \|r_0\|_2 e_1 \tag{8.23}$$

and then evaluating

$$\widetilde{z}_k = R_k^{-1} z_k \quad \text{and} \quad x_k = V_k \widetilde{z}_k. \tag{8.24}$$

We will denote this way of computing by rewriting expression (8.22) with parentheses that indicate the order of computation (in a way that is different from other orders of computation to come):

$$x_k^M = V_k \, (R_k^{-1} Q_{k+1,k}^T \|r_0\|_2 e_1). \tag{8.25}$$

In fact, this is the way of computation that is followed in GMRES for unsymmetric linear systems and therefore we will denote this implementation of the minimum residual approach on the basis of the three-term Lanczos relation as MINRES_{GMRES}.

The disadvantage of MINRES_{GMRES} would be that we have to store all Lanczos vectors and this is avoided in actual implementations. Indeed, the computations in the formula (8.22) can be grouped alternatively as

$$x_k^M = (V_k R_k^{-1}) z_k \equiv W_k z_k, \tag{8.26}$$

with z_k as in (8.23). Because of the banded structure of R_k, it is easy to see that the last column of W_k can be computed from the last two columns of W_{k-1} and v_k. This interpretation makes it possible to generate x_k^M with a short recurrence, since z_k itself follows from the k-th Givens rotation applied to $(z_{k-1}^T, 0)^T$. This approach is the basis for actual MINRES implementations and therefore we will refer to this implementation simply as MINRES.

Note that z_k is evaluated in the same way for MINRES_{GMRES} and MINRES. In fact, the essential difference between the evaluations in (8.24) and (8.26) is in the action of R_k^{-1}. In MINRES_{GMRES} the operator R_k^{-1} acts on z_k, whereas in MINRES it acts on V_k. The analysis in [179] concentrates on this difference.

We start with MINRES_{GMRES}. With the given computed R_k and z_k, we compute in floating point finite precision arithmetic the vector $\widetilde{z}_k$ and we denote the actually computed result as $\widehat{z}_k$. This results satisfies (cf. [98, p.89]):

$$(R_k + \Delta_R)\widehat{z}_k, \quad \text{with} \quad |\Delta_R| \le 3\xi |R_k| + \mathcal{O}(\xi^2),$$

where ξ denotes the machine precision.

Because $\widehat{z}_k = (I + R_k^{-1}\Delta_R)^{-1}R_k^{-1}z_k$, we have for the difference between the computed and the exact vector

$$\Delta_1 \equiv \widehat{z}_k - \widetilde{z}_k = -R_k^{-1}\Delta_R R_k^{-1}z_k.$$

Then we have to multiply V_k with $\widehat{z}_k$ in order to obtain the computed $\widehat{x}_k$, and this also introduces an error (see [116, p.78]):

$$\widehat{x}_k = V_k\widehat{z}_k + \Delta_2, \quad \text{with} \quad |\Delta_2| \leq k\xi|V_k||\widetilde{z}_k| + \mathcal{O}(\xi^2).$$

These two errors Δ_1 and Δ_2 lead to a contribution Δx_k in the computed solution x_k:

$$\Delta x_k = V_k\Delta_1 + \Delta_2.$$

When we evaluate $b - A\widehat{x}_k$ then this contribution leads to an extra deviation Δr_k^G (in addition to other error sources) with respect to the exact $b - Ax_k$, and after some formula manipulation (see [179, p.732]) we obtain

$$\frac{\|\Delta r_k^G\|_2}{\|b\|_2} \leq (3\sqrt{3}\|V_{k+1}\|_2 + k\sqrt{k})\xi\kappa_2(A), \tag{8.27}$$

where $\kappa_2(A)$ denotes the condition number $\|A^{-1}\|_2\|A\|_2$.

In actual situations $\|V_{k+1}\|_2$ is a modest number, much smaller than $\sqrt{k+1}$, because there is at least local orthogonality in the columns of V_{k+1}. The dominating factor is the condition number, but note that we may already expect relative errors in the order of $\xi\kappa_2(A)$ because of rounding errors in the representation of A and b. This leads to the conclusion that the computation of $\widehat{x}_k$ with R_k and the evaluation of $V_k\widehat{z}_k$ is rather stable in MINRES$_{GMRES}$.

Now we turn our attention to the implementation of MINRES. According to (8.26), we have to evaluate W_k. The j-th row $w_{j,:}$ of W_k satisfies

$$w_j R_k = v_{j,:},$$

which in finite precision leads to a row $\widehat{w}_{j,:}$ that satisfies

$$\widehat{w}_{j,:}(R_k + \Delta_{R_j}) = v_{j,:} \quad \text{with} \quad |\Delta_{R_j}| \leq 3\xi|R_k| + \mathcal{O}(\xi^2). \tag{8.28}$$

After combining the relations for the k rows of the computed $\widehat{W}_k$, we have that $\widehat{W}_k$ satisfies

$$\widehat{W}_k = (V_k + \Delta_W)R_k^{-1} \quad \text{with} \quad |\Delta_W| \leq 3\xi|\widehat{W}_k|\,|R_k| + \mathcal{O}(\xi^2). \tag{8.29}$$

In this equation, we may replace $\widehat{W}_k$ in the expression for $|\Delta_W|$ by $V_k R_k^{-1}$, because this leads only to additional errors of $\mathcal{O}(\xi^2)$. The computation of x_k with the computed $\widehat{W}_k$ leads to additional errors that can be expressed as

$$\widehat{x}_k = \widehat{W}_k z_k + \Delta_3 \quad \text{with} \quad |\Delta_3| \leq k\xi |W_k|\, |z_k| + \mathcal{O}(\xi^2).$$

Proceeding in a similar way as for MINRES$_{GMRES}$, we obtain errors Δr_k^M in the computed $b - A\widehat{x}_k$ that can be attributed to the computation of $\widehat{W}_k$ and the assembly of $\widehat{x}_k$, that can be bounded as

$$\frac{\|\Delta r_k^M\|_2}{\|b\|_2} \leq 3\sqrt{3k}\xi\kappa_2(A)^2 + k\sqrt{k}\xi\kappa_2(A). \tag{8.30}$$

Comparing this with the expression (8.27) for MINRES$_{GMRES}$, we see that we now have an error bound for MINRES, due to the evaluation of $\widehat{W}_k$ and the assembly of $\widehat{x}_k$, that is proportional to $\kappa_2(A)^2$. Of course, we might wonder how pessimistic these upperbounds are, but for the example given in Figure 8.5 (taken from [179]) we see that we may really encounter differences in practice that may be explained by the condition numbers.

Our analysis does not imply that MINRES is always an unattractive method. Of course, when A is not ill-conditioned, the method is attractive because of economy in storage. But also, if A is ill-conditioned, then the ill-conditioning is felt in the solution, if that solution has components of about the same magnitude in right singular vector directions corresponding to the smallest and the largest singular vectors. For some classes of problems, including problems from tomography, we want to avoid components in the directions of the small singular vectors (regularization) and in such cases MINRES may be still attractive, because the Krylov methods, including MINRES, tend to discover the small singular vectors after more iterations than the large singular vector directions. This may also be viewed as a sort of regularization, for an analysis of this phenomenon see [193].

8.3.3 Some considerations for CG

The somewhat alarming aspect of our analysis of MINRES and the hypothetical MINRES$_{GMRES}$ is that it gives the impression that we have to pay a high price for short recurrence relations, when we use these to transform the Krylov basis: the transformation from V_k to W_k in MINRES$_{GMRES}$. This may make us slightly nervous with respect to Conjugate Gradients, where the Krylov basis is also transformed, see (5.3): the transformation of the orthogonal R-basis to RU^{-1}. The discussion in [178, Section 4] indicates that the usual CG implementations,

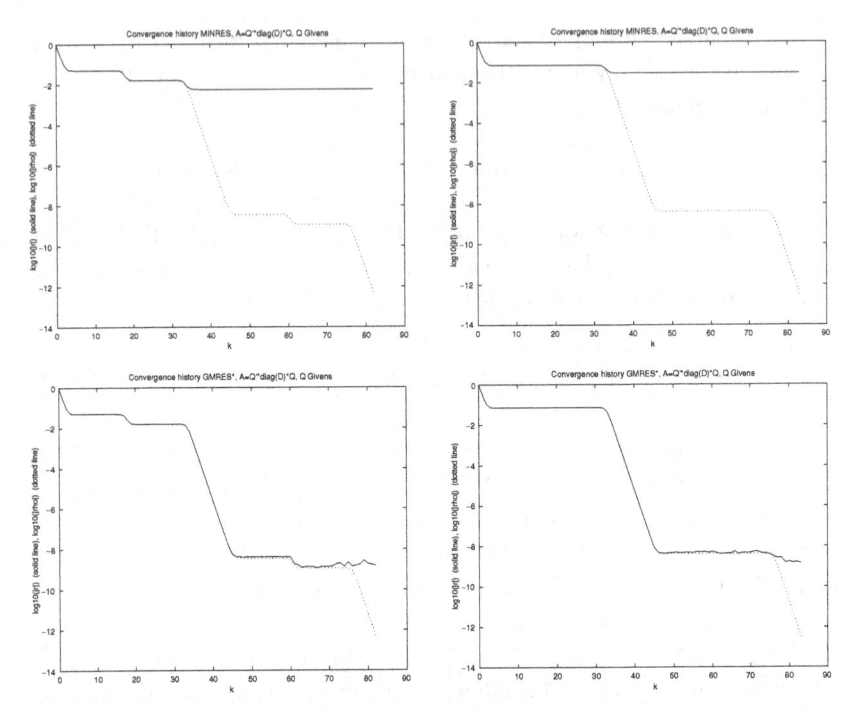

Figure 8.5. MINRES (top) and MINRES_{GMRES} (bottom): solid line (–) $\log_{10}$ of $\|b - A\widehat{x}_k\|_2/\|b\|_2$; dotted line $(\cdots)$ $\log_{10}$ of the estimated residual norm reduction ρ_k. The pictures show the results for a positive definite system (the left pictures) and for an indefinite system (the right pictures). For both examples $\kappa_2(A) = 3 \cdot 10^8$. To be more specific: at the left $A = GDG'$ with D diagonal, $D \equiv \text{diag}(10^{-8}, 2 \cdot 10^{-8}, 2 : h : 3)$, $h = 1/789$, and G the Givens rotation in the $(1, 30)$-plane over an angle of $45°$; at the right $A = GDG'$ with D diagonal $D \equiv \text{diag}(-10^{-8}, 10^{-8}, 2 : h : 3)$, $h = 1/389$, and G the same Givens rotation as for the left example; in both examples b is the vector with all coordinates equal to 1, $x_0 = 0$, and the relative machine precision $\xi = 1.1 \cdot 10^{-16}$.

based on coupled two-term recurrence relations, are very stable and lead to perturbations in the order of $\kappa(A)$ times machine precision at most. The crux is that in MINRES we transform the basis by a three-term recurrence and these three-term recurrence relations may trigger parasytic components in the transformed vectors.

9

Bi-CGSTAB

9.1 A more smoothly converging variant of CGS

We have mentioned that the residuals r_i in CGS satisfy the relation $r_i = P_i(A)^2 r_0$, in which $P_i(A)r_0$ just defines the residual $r_{Bi\text{-}CG,i}$ in the bi-conjugate gradient method:

$$r_{Bi\text{-}CG,i} = P_i(A)r_0.$$

By construction, we have that $(P_i(A)r_0, P_j(A^T)\widehat{r}_0) = 0$ for $j < i$, which expresses the fact that $P_i(A)r_0$ is perpendicular to the subspace $K^i(A^T; \widehat{r}_0)$, spanned by the vectors

$$\widehat{r}_0, A^T\widehat{r}_0, \ldots, (A^T)^{i-1}\widehat{r}_0.$$

This implies that, in principle, we can also recover the bi-conjugate gradient iteration parameters by requiring that, e.g., r_i is perpendicular to $\widetilde{P}_j(P^T)\widehat{r}_0$, or, equivalently, $(\widetilde{P}_j(A)P_i(A)r_0, \widehat{r}_0) = 0$, for another suitable set of polynomials $\widetilde{P}_j$ of degree j. In Bi-CG we take $\widetilde{P}_j = P_j$, namely $\widehat{r}_j = P_j(A^T)\widehat{r}_0$. This is exploited in CG-S, as we have indicated before, since recursion relations for the vectors $P_j^2(A)r_0$ can be derived from those for $P_j(A)r_0$.

Of course, we can now construct iteration methods, by which the x_i are generated so that $r_i = \widetilde{P}_i(A)P_i(A)r_0$ with other i-th degree polynomials, such as, e.g., Chebyshev-polynomials, which might be more suitable. Unfortunately, the optimal parameters for the Chebyshev polynomials are generally not easily obtainable and also the recurrence relations for the resulting method are more complicated than for CG-S. Another possibility is to take for $\widetilde{P}_j$ a polynomial of the form

$$Q_i(x) = (1 - \omega_1 x)(1 - \omega_2 x) \cdots (1 - \omega_i x), \tag{9.1}$$

and to select suitable constants ω_j. This expression leads to an almost trivial recurrence relation for the Q_i.

An obvious possibility for selecting ω_j in the j-th iteration step is to minimize $\|r_j\|$, with respect to ω_j, for residuals that can be written as $r_j = Q_j(A)P_j(A)r_0$. This leads to the Bi-CGSTAB method [201].

We will now derive the (unpreconditioned) Bi-CGSTAB scheme. This will be done completely analogously to the derivation of CGS. We repeat the two important relations for the vectors r_i and p_i, extracted from the unpreconditioned version of Bi-CG, see Figure 7.1 (take $K = I$):

$$\begin{aligned} p_i &= r_{i-1} + \beta_{i-1} p_{i-1} \\ r_i &= r_{i-1} - \alpha_i A p_i. \end{aligned}$$

We write r_i and p_i again in polynomial form:

$$r_i = P_i(A) r_0 \quad \text{and} \quad p_i = T_{i-1}(A) r_0,$$

in which $P_i(A)$ and $T_{i-1}(A)$ are polynomials in A of degree i, $i+1$, respectively. From the Bi-CG recurrence relations we obtain recurrence relations for these polynomials:

$$T_{i-1}(A) r_0 = (P_{i-1}(A) + \beta_{i-1} T_{i-2}(A)) r_0,$$

and

$$P_i(A) r_0 = (P_{i-1}(A) - \alpha_i A T_{i-1}(A)) r_0.$$

In the Bi-CGSTAB scheme we wish to have recurrence relations for

$$\widehat{r}_i = Q_i(A) P_i(A) r_0.$$

With Q_i as in (9.1) and the Bi-CG relation for the factors P_i and T_{i-1}, it then follows that

$$\begin{aligned} Q_i(A) P_i(A) r_0 &= (1 - \omega_i A) Q_{i-1}(A)(P_{i-1}(A) - \alpha_i A T_{i-1}(A)) r_0 \\ &= \{Q_{i-1}(A) P_{i-1}(A) - \alpha_i A Q_{i-1}(A) T_{i-1}(A)\} r_0 \\ &\quad - \omega_i A\{(Q_{i-1}(A) P_{i-1}(A) - \alpha_i A Q_{i-1}(A) T_{i-1}(A))\} r_0. \end{aligned}$$

Clearly, we also need a relation for the product $Q_i(A)T_i(A)r_0$. This can also be obtained from the Bi-CG relations:

$$Q_i(A)T_i(A)r_0 = Q_i(A)(P_i(A) + \beta_i T_{i-1}(A))r_0$$
$$= Q_i(A)P_i(A)r_0 + \beta_i(1 - \omega_i A)Q_{i-1}(A)T_{i-1}(A)r_0$$
$$= Q_i(A)P_i(A)r_0 + \beta_i Q_{i-1}(A)T_{i-1}(A)r_0 - \beta_i\omega_i A Q_{i-1}(A)T_{i-1}(A)r_0.$$

Finally we have to recover the Bi-CG constants ρ_i, β_i, and α_i by inner products in terms of the new vectors that we now have generated. For example, β_i can be computed as follows. First we compute

$$\widetilde{\rho}_i = (\widetilde{r}_0, Q_i(A)P_i(A)r_0) = (Q_i(A^T)\widetilde{r}_0, P_i(A)r_0).$$

By construction, the vector $P_i(A)r_0$ is orthogonal with respect to all vectors $U_{i-1}(A^T)\widetilde{r}_0$, where U_{i-1} is an arbitrary polynomial of degree $\leq i-1$. This means that we have to consider only the highest order term of $Q_i(A^T)$ when computing $\widetilde{\rho}_i$. This term is given by $(-1)^i\omega_1\omega_2\cdots\omega_i(A^T)^i$. We actually wish to compute

$$\rho_i = (P_i(A^T)\widehat{r}_0, P_i(A)r_0),$$

and since the highest order term of $P_i(A^T)$ is given by $(-1)^i\alpha_1\alpha_2\cdots\alpha_i(A^T)^i$, it follows that

$$\beta_i = (\widetilde{\rho}_i/\widetilde{\rho}_{i-1})(\alpha_{i-1}/\omega_{i-1}).$$

The other constants can be derived similarly.

Note that in our discussion we have focused on the recurrence relations for the vectors r_i and p_i, while in fact our main goal is to determine x_i. As in all CG-type methods, x_i itself is not required for continuing the iteration, but it can easily be determined as a 'side product' by realizing that an update of the form $r_i = r_{i-1} - \gamma Ay$ corresponds to an update $x_i = x_{i-1} + \gamma y$ for the current approximated solution.

By writing r_i for $Q_i(A)P_i(A)r_0$ and p_i for $Q_{i-1}(A)T_{i-1}(A)r_0$, we obtain the following scheme for Bi-CGSTAB (I trust that, with the foregoing observations, the reader will now be able to verify the relations in Bi-CGSTAB). In this scheme we have computed the ω_i so that the residual $r_i = Q_i(A)P_i(A)r_0$ is minimized

x_0 is an initial guess; $r_0 = b - Ax_0$
Choose $\widetilde{r}$, for example, $\widehat{r} = r_0$
for $i = 1, 2, \ldots.$
  $\rho_{i-1} = \widetilde{r}^T r_{i-1}$
  if $\rho_{i-1} = 0$ method fails
  if $i = 1$
    $p_i = r_{i-1}$
  else
    $\beta_{i-1} = (\rho_{i-1}/\rho_{i-2})(\alpha_{i-1}/\omega_{i-1})$
    $p_i = r_{i-1} + \beta_{i-1}(p_{i-1} - \omega_{i-1}v_{i-1})$
  endif
  $v_i = Ap_i;$
  $\alpha_i = \rho_{i-1}/\widetilde{r}^T v_i$
  $s = r_{i-1} - \alpha_i v_i$
  check $\|s\|_2$, if small enough: $x_i = x_{i-1} + \alpha_i p_i$ and stop
  $t = As, \omega_i = t^T s/t^T t$
  $x_i = x_{i-1} + \alpha_i p_i + \omega_i s$
  $r_i = s - \omega_i t$
  check convergence; continue if necessary
  for continuation it is necessary that $\omega_i \neq 0$
end

Figure 9.1. The Bi-CGSTAB algorithm.

in 2-norm as a function of ω_i. In Figure 9.1, I have presented the algorithm, skipping the $\widehat{}$ notation for the Bi-CGSTAB iteration vectors.

In order to place a restriction on memory traffic, we have carried out both updates to the current solution x in one single step, while the updates to the residual r had to be done separately ($s = r_{i-1} - \alpha v_i$ and $r_i = s - \omega_i t$). So s represents the residual after a 'Bi-CG step'. If the norm of s is small enough then we might stop, but in that case, before stopping the algorithm, the current solution has to be updated appropriately as $x_i = x_{i-1} + \alpha p_i$ in order to be compatible with the current residual s (and the computation of t, ω_i, as well as the second update $\omega_i s$ should be skipped).

From the orthogonality property $(P_i(A)r_0, Q_j(A^T)\widehat{r}_0) = 0$, for $j < i$, it follows that Bi-CGSTAB is also a finite method, i.e., in exact arithmetic it will terminate after $m \leq n$ iteration steps. In this case we get $s = 0$ at iteration step m and ω_m

is then not defined. This represents a lucky break-down of the algorithm and the process should be terminated as indicated in the previous paragraph.

In the presented form Bi-CGSTAB requires, for the solution of an N by N system $Ax = b$, evaluation of 2 matrix vector products with A, $12N$ flops for vector updates and 4 inner products. This has to be compared with (unpreconditioned) CGS, which also requires 2 matrix vector products with A, and $13N$ flops, but only 2 inner products. In practical situations, however, the two additional inner products lead to only a small increase in computational work per iteration step and this is readily undone by almost any reduction in the number of iteration steps (especially on vector computers for which inner products are usually fast operations).

Except for memory locations for x, b, r, and A we need memory space for 4 additional N-vectors $\widetilde{r}$, p, v, and t for Bi-CGSTAB (note that r may be overwritten by s). This is the same as for CGS.

Of course, Bi-CGSTAB may suffer from the same breakdown problems as Bi-CG and CGS. These problems stem basically from the fact that for general matrices the bilinear form

$$[x, y] \equiv (P(A^T)x, P(A)y),$$

which is used to form the bi-orthogonality, does not define an inner product. In particular, it may occur that, by an unlucky choice for $\widetilde{r}$, an iteration parameter ρ_i or $\widetilde{r}^T v_i$ is zero (or very small), without convergence having taken place. In an actual code we should test for such situations and take appropriate measures, e.g., restart with a different $\widetilde{r}$ or switch to another method (for example GMRES).

The preconditioned Bi-CGSTAB algorithm for solving the linear system $Ax = b$, with preconditioning K, reads as in Figure 9.2.

The matrix K in this scheme represents the preconditioning matrix and the way of preconditioning [201]. The above scheme in fact carries out the Bi-CGSTAB procedure for the explicitly postconditioned linear system

$$AK^{-1}y = b,$$

but the vectors y_i and the residual have been back transformed to the vectors x_i and r_i corresponding to the original system $Ax = b$.

In exact arithmetic, the α_j and β_j have the same values as those generated by Bi-CG and CGS. Hence, they can be used to extract eigenvalue approximations for the eigenvalues of A (see Bi-CG).

```
x_0 is an initial guess, r_0 = b - A x_0
Choose r~, for example, r~ = r_0
for i = 1, 2, ....
    rho_{i-1} = r~^T r_{i-1}
    if rho_{i-1} = 0 method fails
    if i = 1
        p_i = r_{i-1}
    else
        beta_{i-1} = (rho_{i-1}/rho_{i-2})(alpha_{i-1}/omega_{i-1})
        p_i = r_{i-1} + beta_{i-1}(p_{i-1} - omega_{i-1} v_{i-1})
    endif
    Solve p^ from K p^ = p_i
    v_i = A p^
    alpha_i = rho_{i-1} / r~^T v_i
    s = r_{i-1} - alpha_i v_i
    if ||s|| small enough then
        x_i = x_{i-1} + alpha_i p^, quit
    Solve s^ from K s^ = s
    t = A s^
    omega_i = t^T s / t^T t
    x_i = x_{i-1} + alpha_i p^ + omega_i s^
    if x_i is accurate enough then quit
    r_i = s - omega_i t
    for continuation it is necessary that omega_i != 0
end
```

Figure 9.2. The Bi-CGSTAB algorithm with preconditioning.

Bi-CGSTAB can be viewed as the product of Bi-CG and GMRES(1). Of course, other product methods can also be formulated. Gutknecht [109] has proposed BiCGSTAB2, which is constructed as the product of Bi-CG and GMRES(2).

9.2 Bi-CGSTAB(2) and variants

One particular weak point in Bi-CGSTAB is that we get breakdown if an ω_j is equal to zero. We may equally expect negative effects when ω_j is small. In fact, Bi-CGSTAB can be viewed as the combined effect of Bi-CG and GCR(1), or GMRES(1), steps. As soon as the GCR(1) part of the algorithm (nearly) stagnates, then the Bi-CG part in the next iteration step cannot (or can only poorly) be constructed.

Another dubious aspect of Bi-CGSTAB is that the factor Q_k has only real roots by construction. It is well known that optimal reduction polynomials for matrices with complex eigenvalues may also have complex roots. If, for instance, the matrix A is real skew-symmetric, then GCR(1) stagnates forever, whereas a method like GCR(2) (or GMRES(2)), in which we minimize over two combined successive search directions, may lead to convergence, and this is mainly due to the fact that the complex eigenvalue components in the error can be effectively reduced.

This point of view was taken in [109] for the construction of a variant called Bi-CGSTAB2. In the odd-numbered iteration steps the Q-polynomial is expanded by a linear factor, as in Bi-CGSTAB, but in the even-numbered steps this linear factor is discarded, and the Q-polynomial from the previous even-numbered step is expanded by a quadratic $1 - \alpha_k A - \beta_k A^2$. For this construction the information from the odd-numbered step is required. It was anticipated that the introduction of quadratic factors in Q might help to improve convergence for systems with complex eigenvalues, and, indeed, some improvement was observed in practical situations (see also [157]).

However, my presentation suggests a possible weakness in the construction of Bi-CGSTAB2, namely in the odd-numbered steps the same problems may occur as in Bi-CGSTAB. Since the even-numbered steps rely on the results of the odd-numbered steps, this may equally lead to unnecessary breakdowns or poor convergence. In [174] another, and even simpler, approach was taken to arrive at the desired even-numbered steps, without the necessity of the construction of the intermediate Bi-CGSTAB-type step in the odd-numbered steps. Hence, in this approach the polynomial Q is constructed straightaway as a product of quadratic factors, without ever constructing a linear factor. As a result the new method Bi-CGSTAB(2) leads only to significant residuals in the even-numbered steps and the odd-numbered steps do not lead necessarily to useful approximations.

In fact, it is shown in [174] that the polynomial Q can also be constructed as the product of ℓ-degree factors, without the construction of the intermediate lower degree factors. The main idea is that ℓ successive Bi-CG steps are carried out, where for the sake of an A^T-free construction the already available part of Q is expanded by simple powers of A. This means that after the Bi-CG part of the algorithm vectors from the Krylov subspace $s, As, A^2s, \ldots, A^\ell s$, with $s = P_k(A)Q_{k-\ell}(A)r_0$, are available, and it is then relatively easy to minimize the residual over that particular Krylov subspace. There are variants of this approach in which more stable bases for the Krylov subspaces are generated

[177], but for low values of ℓ a standard basis satisfies, together with a minimum norm solution obtained through solving the associated normal equations (which requires the solution of an ℓ by ℓ system). In most cases Bi-CGSTAB(2) will already give nice results for problems where Bi-CGSTAB or Bi-CGSTAB2 may fail. Note, however, that, in exact arithmetic, if no breakdown situation occurs, Bi-CGSTAB2 would produce exactly the same results as Bi-CGSTAB(2) at the even-numbered steps.

Bi-CGSTAB(2) can be represented by the scheme in Figure 9.3. For more general Bi-CGSTAB(ℓ) schemes see [174, 177].

x_0 is an initial guess, $r_0 = b - Ax_0$
Choose $\widehat{r}_0$, for example $\widehat{r}_0 = r$
$\rho_0 = 1, u = 0, \alpha = 0, \omega_2 = 1$
for $i = 0, 2, 4, 6, \ldots.$

$\rho_0 = -\omega_2 \rho_0$

even Bi-CG step: $\rho_1 = (\widehat{r}_0, r_i), \beta = \alpha\rho_1/\rho_0, \rho_0 = \rho_1$

$u = r_i - \beta u$

$v = Au$

$\gamma = (v, \widehat{r}_0), \alpha = \rho_0/\gamma$

$r = r_i - \alpha v$

$s = Ar$

$x = x_i + \alpha u$

odd Bi-CG step: $\rho_1 = (\widehat{r}_0, s), \beta = \alpha\rho_1/\rho_0, \rho_0 = \rho_1$

$v = s - \beta v$

$w = Av$

$\gamma = (w, \widehat{r}_0), \alpha = \rho_0/\gamma$

$u = r - \beta u$

$r = r - \alpha v$

$s = s - \alpha w$

$t = As$

GCR(2)-part: $\omega_1 = (r, s), \mu = (s, s), \nu = (s, t), \tau = (t, t)$

$\omega_2 = (r, t), \tau = \tau - \nu^2/\mu, \omega_2 = (\omega_2 - \nu\omega_1/\mu)/\tau$

$\omega_1 = (\omega_1 - \nu\omega_2)/\mu$

$x_{i+2} = x + \omega_1 r + \omega_2 s + \alpha u$

$r_{i+2} = r - \omega_1 s - \omega_2 t$

if x_{i+2} is accurate enough **then** quit

$u = u - \omega_1 v - \omega_2 w$

end

Figure 9.3. The Bi-CGSTAB(2) algorithm.

Another advantage of Bi-CGSTAB(2) over BiCGSTAB2 is its efficiency. The Bi-CGSTAB(2) algorithm requires 14 vector updates, 9 inner products and 4 matrix vector products per full cycle. This has to be compared with a combined odd-numbered and even-numbered step in BiCGSTAB2, which requires 22 vector updates, 11 inner products, and 4 matrix vector products, and with two steps of Bi-CGSTAB, which require 4 matrix vector products, 8 inner products and 12 vector updates. The numbers for BiCGSTAB2 are based on an implementation described in [157].

Also with respect to memory requirements, Bi-CGSTAB(2) takes an intermediate position: it requires 2 n-vectors more than Bi-CGSTAB and 2 n-vectors less than Bi-CGSTAB2.

For distributed memory machines the inner products may cause communication overhead problems (see, e.g., [48]). We note that the Bi-CG steps are very similar to conjugate gradient iteration steps, so that we may consider all kind of tricks that have been suggested to reduce the number of synchronization points caused by the 4 inner products in the Bi-CG parts. For an overview of these approaches see [20]. If on a specific computer it is possible to overlap communication with communication, then the Bi-CG parts can be rescheduled so as to create overlap possibilities:

(1) The computation of ρ_1 in the even Bi-CG step may be done just before the update of u at the end of the GCR part.
(2) The update of x_{i+2} may be delayed until after the computation of γ in the even Bi-CG step.
(3) The computation of ρ_1 for the odd Bi-CG step can be done just before the update for x at the end of the even Bi-CG step.
(4) The computation of γ in the odd Bi-CG step already has overlap possibilities with the update for u.

For the GCR(2) part we note that the 5 inner products can be taken together, in order to reduce start-up times for their global assembling. This gives the method Bi-CGSTAB(2) a (slight) advantage over Bi-CGSTAB. Furthermore we note that the updates in the GCR(2) may lead to more efficient code than for Bi-CGSTAB, since some of them can be combined.

9.3 More general hybrid Bi-CG methods

As follows from Sonneveld's paper [180], Bi-CG can be the basis for methods that generate residuals of the form

$$r_i = Q_i(A)P_i(A)r_0,$$

where $P_i(A)$ is the iteration polynomial defined by the Bi-CG recursions. The trick is to find suitable, easy to generate, polynomials $Q_i(A)$ that help to decrease the norm of r_i. In CGS we select $Q_i = P_i$, in Bi-CGSTAB we select Q_i as the product of GMRES(1) iteration polynomials, and in Bi-CGSTAB(ℓ) the polynomial Q_i is a product of GMRES(ℓ) polynomials for i a multiple of ℓ.

Zhang [225] has studied the case where Q_i is also generated by a three term recurrence relation similar to the underlying Lanczos recurrence relation (9.2) of Bi-CG.

Exercise 9.1. *Show, similarly to the derivation of relation (5.7) for CG, that the iteration polynomial P_i of Bi-CG satisfies*

$$P_i(A) = (1 + \frac{\beta_{i-2}}{\alpha_{i-2}} - \alpha_{i-1}A)P_{i-1}(A) - \frac{\beta_{i-2}}{\alpha_{i-2}}P_{i-2}(A), \tag{9.2}$$

for $i \geq 2$. How are $P_0(A)$ and $P_1(A)$ defined?

Equation (9.2) inspired Zhang to exploit a similar relation for Q_i:

$$Q_i(A) = (1 + \eta_{i-1} - \zeta_{i-1}A)Q_{i-1}(A) - \zeta_{i-1}Q_{i-2}(A). \tag{9.3}$$

The combination of (9.3) with $P_i(A)$ leads to the following expression for $r_i = Q_i(A)P_i(A)r_0$:

$$\begin{aligned} Q_i(A)P_i(A)r_0 = {} & Q_{i-1}P_i(A)r_0 - \eta_{i-1}(Q_{i-2}(A) - Q_{i-1}(A))P_i(A)r_0 \\ & - \zeta_{i-1}AQ_{i-1}P_i(A)r_0. \end{aligned} \tag{9.4}$$

Similarly to the derivation of Bi-CGSTAB (and CGS) we can, exploiting the recursions for the polynomials P_i and T_i, defined by the Bi-CG algorithm as the generating functions for the residuals and update vectors, respectively, derive recursions for each of the product forms at the right-hand side of (9.4). We denote these products as

$$t_{i-1} = Q_{i-1}(A)P_i(A)r_0 \tag{9.5}$$

$$y_{i-1} = (Q_{i-2}(A) - Q_{i-1}(A))P_i(A)r_0, \tag{9.6}$$

which leads to the following recursion for r_i:

$$r_i = t_{i-1} - \eta_{i-1}y_{i-1} - \zeta_{i-1}At_{i-1}. \tag{9.7}$$

Zhang [225, Section 5.3] suggests determining the parameters η_{i-1} and ζ_{i-1} so that they minimize the norm $\|r_i\|_2$ as a function of these parameters. The resulting algorithm is called GPBi-CG (generalized product Bi-CG) [225].

It can be shown that if we take $\eta_i = 0$ for even i then GPBi-CG reduces to Bi-CGSTAB(2) (in exact arithmetic). It is tempting to believe that GPBi-CG is superior to Bi-CGSTAB(2), because it minimizes, locally, the new contribution to the Q_i factor. However, it is still unclear whether the complete Q_i polynomial in GPBi-CG should have better reduction properties than the product of the GMRES(2) polynomials in Bi-CGSTAB(2). Moreover, the Q_i polynomial in

x_0 is an initial guess; $r_0 = b - Ax_0$
$u = z = 0$
Choose $\widetilde{r}$, for example, $\widetilde{r} = r_0$
for $i = 1, 2, \ldots.$
  $\rho_{i-1} = \widetilde{r}^H r_{i-1}$
  if $\rho_{i-1} = 0$ method fails
  if $i = 1$
    $p = r_{i-1}, q = Ap$
    $\alpha_i = \rho_{i-1}/\widetilde{r}^H q$
    $t = r_{i-1} - \alpha_i q, v = At,$
    $y = \alpha_i q - r_{i-1}$
    $\mu_2 = v^H t; \mu_5 = v^H v, \zeta = \mu_2/\mu_5, \eta = 0$
  else
    $\beta_{i-1} = (\rho_{i-1}/\rho_{i-2})(\alpha_{i-1}/\zeta)$
    $w = v + \beta_{i-1} q$
    $p = r_{i-1} + \beta_{i-1}(p - u), q = Ap$
    $\alpha_i = \rho_{i-1}/\widetilde{r}^H q, s = t - ri - 1$
    $t = r_{i-1} - \alpha_i q, v = At$
    $y = s - \alpha_i(w - q)$
    $\mu_1 = y^H y, \mu_2 = v^H t, \mu_3 = y^H t, \mu_4 = v^H y, \mu_5 = v^H v$
    $\tau = \mu_5\mu_1 - \bar{\mu}_4\mu_4, \zeta = (\mu_1\mu_2 - \mu_3\mu_4)/\tau, \eta = (\mu_5\mu_3 - \bar{\mu}_4\mu_2)/\tau$
  endif
  $u = \zeta q + \eta(s + \beta_{i-1} u)$
  $z = \zeta r_{i-1} + \eta z - \alpha_i u$
  $x_i = x_{i-1} + \alpha_i p + z$
  if x_i is accurate enough **then** quit
  $r_i = t - \eta y - \zeta v$
  for continuation it is necessary that $\zeta \neq 0$
end

Figure 9.4. The GPBi-CG algorithm without preconditioning.

GPBi-CG is constructed by a three term recurrence relation, instead of the two coupled two term recurrences for the P_i polynomial. Three term recurrences are considered as less numerically stable than coupled two term recurrences, so the numerical stability of GPBi-CG also requires further investigation.

Exercise 9.2. *Show that GPBi-CG reduces formally to Bi-CGSTAB (that is, it generates the same approximations in exact arithmetic) if we set $\eta_i = 0$ for all i.*

For completeness, we give the complete GPBi-CG algorithm [225, Algorithm 5] in Figure 9.4. This is also the basis of the Matlab code with which we have carried out some numerical experiments. This algorithm is without preconditioning, but it is straightforward to include preconditioning, for instance, by replacing A by the preconditioned operator $K^{-1}A$ and b by $K^{-1}b$.

Exercise 9.3. *Compare the numbers of floating point operations, in terms of matrix vector operations, inner products, and vector updates, for CGS, Bi-CGSTAB, and GPBi-CG.*

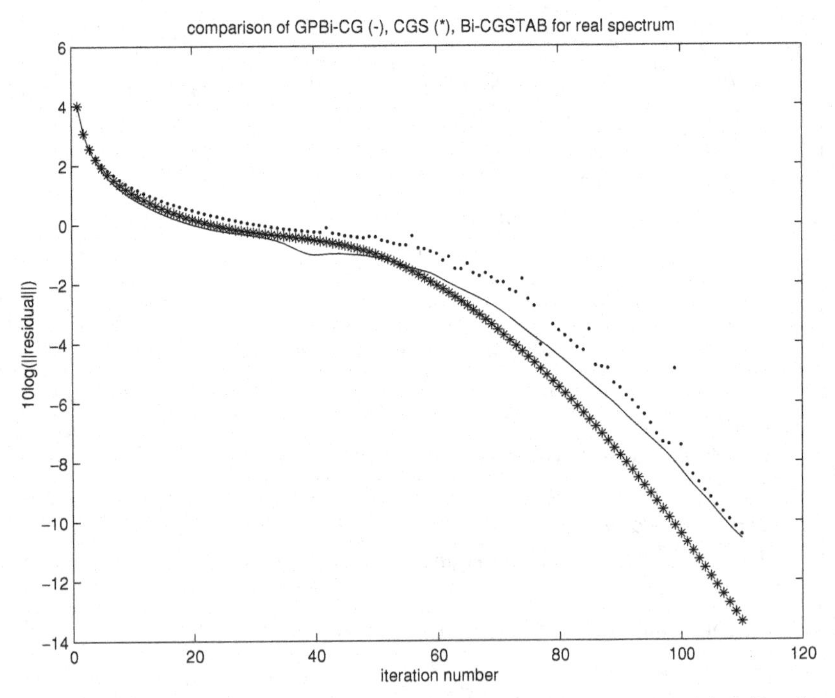

Figure 9.5. GPBi-CG (–), CGS (*) , and Bi-CGSTAB (..) for a system with uniformly distributed eigenvalues.

9.3.1 Numerical experiments

We start with a very simple example. The 300 by 300 matrix A is diagonal, with diagonal elements uniformly distributed in the interval $[1, 1000]$. The right-hand side is chosen so that the solution is the vector of all 1s. In Figure 9.5 we see the convergence history for the three methods, CGS, Bi-CGSTAB, and GPBi-CG. We see that for such nice spectra the methods require about the same number of iterations, with CGS having a slight advantage.

Now we replace the diagonal part from 296 to 299 by two 2 by 2 blocks with eigenvalues $0.1 \pm 2i$, $0.4 \pm i$, respectively. The result for this mildly complex spectrum (4 complex conjugate eigenvalues at the lower end of the spectrum) is shown in Figure 9.6.

In this case CGS starts to show its characteristic peaks: the large peak of about 10^{10} for the residual norm destroys the accuracy of the solution, which has about 5 decimals of accuracy. The solutions obtained with GPBi-CG and Bi-CGSTAB are much more accurate and these methods are, for high accuracy, also slightly faster (in terms of iteration counts). Note that the iterations

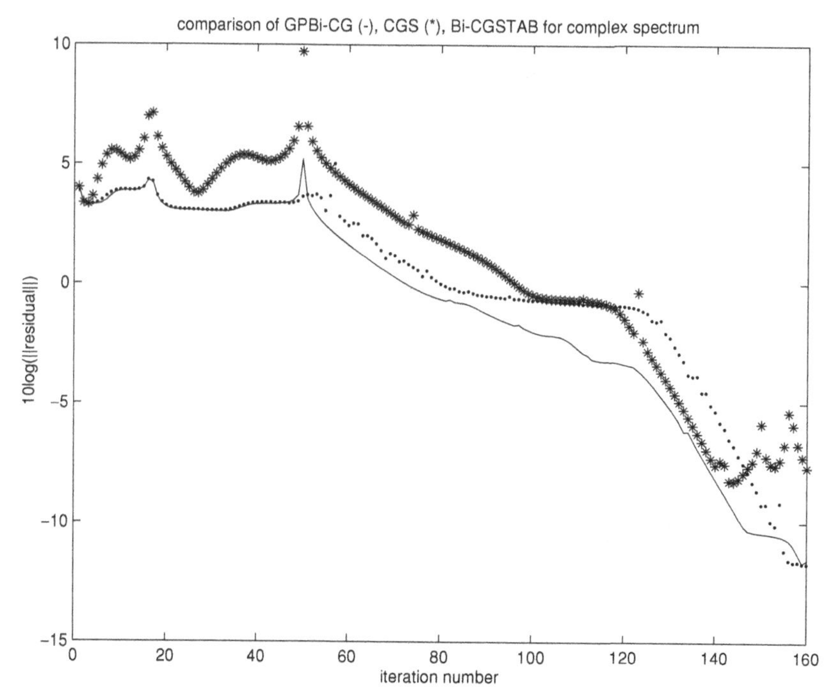

Figure 9.6. GPBi-CG (–), CGS (*), and Bi-CGSTAB (..) for a system with uniformly distributed real eigenvalues and 4 complex ones.

with GPBi-CG are more expensive than those of Bi-CGSTAB. Zhang [225] reports experiments with Helmholtz systems, where GPBi-CG is considerably faster than Bi-CGSTAB(2) and Bi-CGSTAB, and much more accurate than CGS. It seems that in those experiments Bi-CGSTAB suffers from stagnation because of small reductions in its GMRES(1) part. In such cases GPBi-CG and Bi-CGSTAB(ℓ) may be good alternatives, and in particular GPBi-CG deserves more attention by the research community than it has received.

10

Solution of singular systems

10.1 Only nonzero eigenvalues matter

Suppose that we want to solve the linear system $Ax = b$, with a singular n by n matrix A. This can be done with Krylov subspace methods, as I will now show. For simplicity, I will assume that A is symmetric.

I will assume that the system is consistent, that is b has no component in the kernel of A (if the system is inconsistent then the linear system could still be solved with a minimum residual approach). We denote the eigenvalues and eigenvectors of A by λ_j and v_j:

$$Av_j = \lambda_j v_j.$$

Assuming that A has a complete eigensystem then, in principle, the vector b can be expressed in terms of these eigenvectors. We write b as

$$b = \sum_{j=1}^{m} \gamma_j w_j,$$

where m is the number of *different* eigenvalues, counted as $\lambda_1, \ldots, \lambda_m$, for which b has a nonzero component in the corresponding eigenvector direction(s). If λ_j is an eigenvalue with multiplicity 1 then $w_j = v_j$; if λ_j is multiple, then w_j represents the sum of the corresponding eigenvectors.

Now suppose, for simplicity, that for a proper Krylov subspace method the starting vector is $x_0 = 0$, so that the Krylov subspaces are generated with A and b. We will restrict ourselves to the Ritz–Galerkin approach, so that the residual r_i is orthogonal to the i dimensional Krylov subspace $\mathcal{K}^i(A; b)$. For a Lanczos method, we then have that the Lanczos vectors are multiples of the residuals r_i (see, for instance, Section 5.1). These residuals can be expressed as (cf. (5.12))

$$r_i = P_i(A)r_0 = \sum_{j=1}^{m} \gamma_j P_i(\lambda_j)w_j,$$

with P_i an i-th degree polynomial with $P_i(0) = 0$. From this it follows that

$$\begin{aligned} r_\ell = 0 \;\leftrightarrow\; & \sum_{j=1}^{m} \gamma_j P_\ell(\lambda_j)w_j = 0 \\ \leftrightarrow\; & P_\ell(\lambda_j) = 0 \;\text{ for }\; j = 1, ..., m \end{aligned} \tag{10.1}$$

From this it follows that $r_\ell = 0$ if and only if $\ell = m$, that is the construction of the Lanczos basis terminates after exactly m steps. The reduced matrix $T_{m,m}$ has eigenvalues $\lambda_1, \ldots, \lambda_m$ and hence it is nonsingular and can be used to compute the exact solution x, with no components in the kernel of A, as (cf. (5.3))

$$x = x_m = R_m T_{m,m}^{-1} e_1.$$

Moreover, if all λ_j (for which $\lambda_j \neq 0$) are positive, then $T_{i,i}$ is positive definite for all $i \leq m$. This implies that we can safely use the CG method to generate approximations r_i, according to (5.3).

If these λ_js are not all positive, then the Ritz values are in the interval spanned by the smallest and largest of these eigenvalues, and a Ritz value may occasionally for some i be close to 0, or even be equal to 0. Because the iteration polynomial $P_i(0) = 1$, this could imply that r_i is large, or not defined (similar to the usual behaviour of Krylov methods for nonsingular indefinite systems). In such cases we can still construct a Krylov basis, but we have to be careful when solving the reduced system. Of course, we can then solve the reduced system in a minimum residual way, by MINRES, or use SYMMLQ.

If A is unsymmetric and singular, then we can follow the same arguments and use, for instance, GMRES as an iterative process. For our theoretical considerations we then have to use a Jordan decomposition of A, similarly to the convergence analysis in [206]. Also, other Krylov methods can be used, with the same risks as for indefinite matrices, if the matrix restricted to the subspace for which the components $\gamma_j \neq 0$ is indefinite.

10.2 Pure Neumann problems

Singular consistent linear systems arise after discretization of, for instance, a partial differential equation such as

$$-u_{xx} - u_{yy} = f(x, y), \tag{10.2}$$

over a domain Γ with Neumann boundary condition

$$\frac{\partial u}{\partial n} = 0,$$

along the entire boundary $\delta\Gamma$. This means that the solution of (10.2) is determined up to a constant and this is, after proper discretization, reflected by singularity of the matrix A in the discretized system $Ax = b$. The vector with all 1s is then an eigenvector of A with eigenvalue 0. If the linear system $Ax = b$ is consistent, in this case if b is orthogonal to the eigenvector corresponding to the zero eigenvalue, then we may solve the system with CG without further complication, as we have seen in the previous section.

However, it might seem a good idea to fix the solution u at one boundary point and this leads to a nonsingular problem and we might expect to be better off. This is not always a good idea as is shown by the next example.

A numerical example Consider the problem (10.2) over the unit square with an equidistant rectangular grid with mesh spacing $h = 1/19$. This leads, with the usual 5-point finite volume discretization (see [212, Section 6.2]), and after multiplication with h, to a linear system $A_0x = b$. We have used the index 0 in order to underline the singularity of the system. The stencil for A_0 for an interior point in the grid is

$$\begin{array}{ccc} & -1 & \\ -1 & 4 & -1 \\ & -1 & \end{array},$$

while for points on the boundary the stencil has no connections pointing outward of Γ and the central element 4 is then reduced by 1 for each missing connection. The matrix A_0 is of order 400 and has a block structure with blocks of dimension 20.

For our experiment we need a consistent system and we construct a meaningful right-hand side b as follows. We take the vector w with elements $w_j = \sin(j)$ and we eliminate the component in the direction of the vector with all 1s. The resulting vector b is then orthogonal to the vector with all 1s (it is easy to see that that vector is the eigenvector of A_0 with eigenvalue 0).

The system can be solved with CG, because the effective eigenvalues are all positive. The condition number defined by these positive eigenvalues is $\kappa_{effect} \approx 322.9$. In Figure 10.1 we see the convergence history in terms of the 2-norms of the residuals, represented by the solid line.

Now we fix the singularity, by prescribing one value of u at the boundary $\delta\Gamma$: in our experiment we set $b_{60} = 1$ and we set all elements of A_0, except

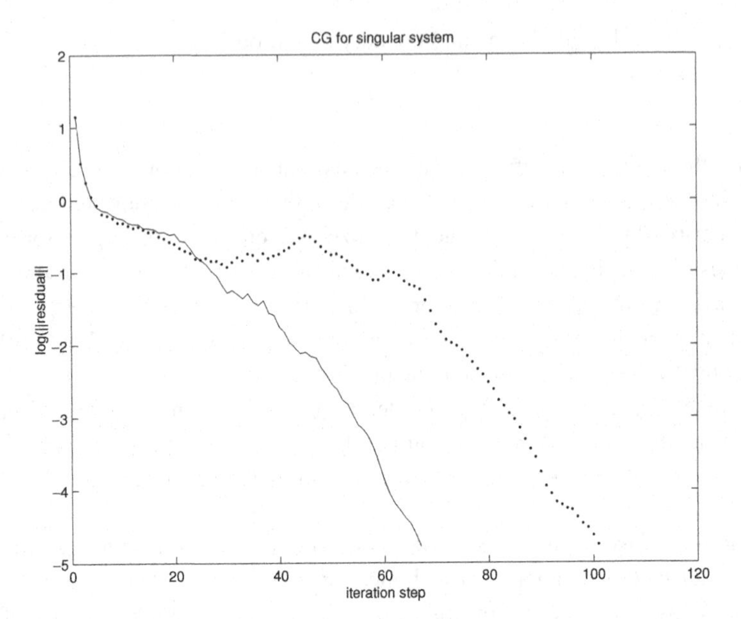

Figure 10.1. CG for a singular (—) and a near-singular (...) systems.

the diagonal element, in the 60-th row and 60-th column to zero. The element $A_0(60, 60) = 1$. It can be shown that the resulting system $A_1 x = \bar{b}$ is nonsingular (using the fact that A_1 is irreducibly diagonally dominant and [212, Theorem 1.8]). The convergence behaviour of CG for this system $A_1 x = \bar{b}$ is represented by the dots in Figure 10.1. Note that we have a noticeably slower convergence now, which is explained by the condition number of A_1, which is $\kappa_2(A_1) \approx 4183.6$. Fixing the nonsingularity of A_0 has resulted in this case in a shift of the eigenvalue 0 to a small nonzero eigenvalue of A_1 that is about 13 times smaller than the smallest nonzero eigenvalue of A_0.

In [125, Section 4], the situation for mildly inconsistent linear systems is analysed. In that case the norms of the residuals tend to increase after an initial phase of convergence. This divergence can be avoided by keeping the iteration vectors in CG explicitly orthogonal with respect to the null space of A. Figure 1 in [124] seems to indicate that the divergence of the residual norms only sets in after a level of ϵ has been reached, where ϵ is the norm of the deviation of b to the nearest right-hand side that makes the system consistent. If this is true in general, then it leads to the conclusion that reorthogonalization with respect to the null space of A is not necessary for consistent singular linear systems.

11

Solution of $f(A)x = b$ with Krylov subspace information

11.1 Introduction

In this chapter, I expand on an idea for exploiting Krylov subspace information obtained for the matrix A and the vector b. This subspace information can be used for the approximate solution of a linear system $f(A)x = b$, where f is some analytic function, $A \in \mathbb{R}^{n\times n}$, and $b \in \mathbb{R}^n$. I will make suggestions on how to use this for the case where f is the matrix *sign* function. The matrix *sign* function plays an important role in QCD computations, see for instance [147].

In [197] an approach was suggested for the use of a Krylov subspace for the computation of approximate solutions of linear systems

$$f(A)x = b.$$

The approach was motivated by the function $f(A) = A^2$, which plays a role in the solution of some biharmonic systems. The approach is easily generalized for nonsymmetric complex matrices, but we may have to pay more attention to the evaluation of f for the reduced system, associated with the Krylov subspace.

In particular, I will discuss some possible approaches in which the Krylov subspace is used for the computation of $\text{sign}(A)p$ for given vectors p. With the evaluation of the matrix *sign* function we have to be extremely careful. A popular approach, based on a Newton iteration, converges fast, but is sensitive for rounding errors, especially when A is ill-conditioned. We will briefly discuss a computational method that was suggested (and analysed) by Bai and Demmel [15]. This approach can also be combined, in principle, with the subspace reduction technique.

11.2 Reduced systems

With equation (3.18) we can construct approximate solutions for $Ax = b$ in the Krylov subspace $K^m(A; r_0)$. These approximate solutions can be written as $x_m = x_0 + V_m y$, with $y \in \mathbb{R}^n$, since the columns of V_m span a basis for the Krylov subspace. The Ritz–Galerkin orthogonality condition for the residual leads to

$$b - Ax_m \perp \{v_1, \ldots, v_m\},$$

or

$$V_m^H (b - A(x_0 + V_m y)) = 0.$$

Now we use $b - Ax_0 = r_0 = \|r_0\|_2 v_1$, and with (3.18) we obtain

$$H_{m,m} y = \|r_0\| e_1, \tag{11.1}$$

with e_1 the first canonical basis vector in $\mathbb{R}^m$. If $H_{m,m}$ is not singular then we can write the approximate solution x_m as

$$x_m = \|r_0\|_2 V_m H_{m,m}^{-1} e_1. \tag{11.2}$$

Note that this expression closely resembles the expression $x = A^{-1}b$ for the exact solution of $Ax = b$. The matrix $H_{m,m}$ can be interpreted as the restriction of A with respect to $v_1, \ldots, v_m$. The vector $\|r_0\|e_1$ is the expression for the right-hand side with respect to this basis, and V_m is the operator that expresses the solution of the reduced system (in $\mathbb{R}^m$) in terms of the canonical basis for $\mathbb{R}^n$.

From now on we will assume, without loss of generality, that $x_0 = 0$. We can also use the above mechanism for the solution of more complicated systems of equations. Suppose that we want to find approximate solutions for $A^2x = b$, with only the Krylov subspace for A and $r_0 = b$ available. The solution of $A^2x = b$ can be realized in two steps

(1) Solve z_m from $Az = b$, using the Ritz–Galerkin condition. With $z = V_m y$ and (3.18), we have that

$$z = \|b\|_2 V_m H_{m,m}^{-1} e_1.$$

(2) Solve x_m from $Ax_m = z_m$, with $x_m = V_m u$. It follows that

$$AV_m u = \|b\|_2 V_m H_{m,m}^{-1} e_1,$$

$$V_{m+1} H_{m+1,m} u_m = \|b\|_2 V_m h_{m,m}^{-1}.$$

The Ritz–Galerkin condition with respect to V_m leads to

$$H_{m,m}u_m = \|b\|_2 H_{m,m}^{-1} e_1.$$

These two steps lead to the approximate solution

$$x_m = \|b\|_2 V_m H_{m,m}^{-2} e_1. \tag{11.3}$$

If we compare (11.2) for $Ax = b$ with (11.3) for $A^2 x = b$, then we see that the operation with A^2 translates to an operation with $H_{m,m}^2$ for the reduced system and that is all.

Note that this approximate solution, x_m, does not satisfy a Ritz–Galerkin condition for the system $A^2 x = b$. Indeed, for $x_m = V_m y$, we have that

$$A^2 V_m y = A V_{m+1} H_{m+1,m} y = V_{m+2} H_{m+2,m+1} H_{m+1,m} y.$$

The Ritz–Galerkin condition with respect to V_m, for $b - Ax_m$, leads to

$$V_m^H V_{m+2} H_{m+2,m+1} H_{m+1,m} y = \|b\|_2 e_1.$$

A straightforward evaluation of $H_{m+2,m+1} H_{m+1,m}$ and the orthogonality of the v_js, leads to

$$V_m^H V_{m+2} H_{m+2,m+1} H_{m+1,m} = H_{m,m}^2 + h_{m+1,m} h_{m,m+1} e_m e_m^T.$$

This means that the reduced matrix for A^2, expressed with respect to the V_m basis, is given by the matrix $H_{m,m}^2$ in which the bottom right element $h_{m,m}$ is updated with $h_{m+1,m} h_{m,m+1}$. By computing x_m as in (11.3), we have ignored the factor $h_{m+1,m} h_{m,m+1}$. This is acceptable, since in generic situations the convergence of the Krylov solution process for $Ax = b$ goes hand in hand with small elements $h_{m+1,m}$.

We can go one step further, and try to solve

$$(A^2 + \alpha A + \beta I)x = b, \tag{11.4}$$

with Krylov subspace information obtained for $Ax = b$ (with $x_0 = 0$). The Krylov subspace $K^m(A, r_0)$ is shift invariant, that is

$$K^m(A, r_0) = K^m(A - \sigma I, r_0),$$

for any scalar $\sigma \in \mathbb{C}$. The matrix polynomial $A^2 + \alpha A + \beta I$ can be factored into

$$A^2 + \alpha A + \beta I = (A - \omega_1 I)(A - \omega_2 I).$$

This can be used to obtain an approximate solution similarly to that for the system $A^2x = b$, see the next exercise.

Exercise 11.1. *Proceed as for A^2, that is solve the given system in two steps, and impose a Ritz–Galerkin condition for each step. Show that this leads to the approximate solution*

$$x_m = \|b\|_2 V_m (H_{m,m}^2 + \alpha H_{m,m} + \beta I_m)^{-1} e_1.$$

for the linear system (11.4).

The generalization to higher degree polynomial systems

$$p_n(A)x \equiv \left(A^n + \alpha_{n-1}A^{n-1} + \cdots + \alpha_0 I\right) x = b$$

is straightforward and leads to an approximate solution of the form

$$x_m = \|b\|_2 V_m p_n(H_{m,m})^{-1} e_1.$$

If f is an analytic function, then we can compute the following approximate solution x_m for the solution of $f(A)x = b$:

$$x_m = \|b\|_2 V_m f(H_{m,m})^{-1} e_1. \tag{11.5}$$

All these approximations are equal to the exact solution if $h_{m+1,m} = 0$. Because $h_{n+1,n} = 0$, we have the exact solution after at most n iterations. The hope is, of course, that the approximate solutions are sufficiently good after $m \ll n$ iterations. There is little to control the residual for the approximate solutions, since in general $f(A)$ may be an expensive function. We use the Krylov subspace reduction in order to avoid expensive evaluation of $f(A)p$ for $p \in \mathbb{R}^n$. It is possible to compare successive approximations x_m and to base a stopping criterion on this comparison.

11.3 Computation of the inverse of $f(H_{m,m})$

The obvious way of computing $f(H_{m,m})^{-1}$ is to reduce $H_{m,m}$ first to some convenient canonical form, for instance to diagonal form. If $H_{m,m}$ is symmetric (in that case $H_{m,m}$ is tridiagonal) then it can be orthogonally transformed to diagonal form:

$$Q_m^H H_{m,m} Q_m = D_m,$$

with Q_m an m by m orthogonal matrix and D_m a diagonal matrix. We then have that

$$Q^H f(H_{m,m})^{-1} Q = f(D_m)^{-1},$$

and this can be used for an efficient and stable computation of x_m. If $H_{m,m}$ is neither symmetric nor (close to) normal (that is $H_{m,m}^H H_{m,m} = H_{m,m} H_{m,m}^H$), then the transformation to diagonal form cannot be done by an orthogonal operator. If $H_{m,m}$ has no Jordan blocks, the transformation can be done by

$$X_m^{-1} H_{m,m} X_m = D_m.$$

This decomposition is not advisable if the condition number of X_m is much larger than 1. In that case it is much better to reduce the matrix $H_{m,m}$ to Schur form:

$$Q_m^H H_{m,m} Q_m = U_m,$$

with U_m an upper triangular matrix. The eigenvalues of $H_{m,m}$ appear along the diagonal of U_m. If A is real, then the computations can be kept in real arithmetic if we use the property that $H_{m,m}$ can be orthogonally transformed to generalized Schur form. In a generalized Schur form, the matrix U_m may have two by two nonzero blocks along the diagonal (but its strict lower triangular part is otherwise zero). These two by two blocks represent complex conjugate eigenpairs of $H_{m,m}$. For further details on Schur forms, generalized Schur forms, and their computation see [98].

11.4 Numerical examples

My numerical examples have been taken from [197]. These experiments have been carried out for diagonal real matrices A, which does not mean a loss of generality (cf. Section 5.2).

The diagonal matrix A is of order 900. Its eigenvalues are 0.034, 0.082, 0.127, 0.155, 0.190. The remaining 895 eigenvalues are uniformly distributed over the interval [0.2, 1.2]. This type of eigenvalue distribution is more or less what we might get with preconditioned Poisson operators. Now suppose that we want to solve $A^2 x = b$, with b a vector with all 1s. We list the results for two different approaches:

(A) Generate the Krylov subspace with A and b and compute the approximate solution x_m^{new} as in (11.3).

Table 11.1. Residual norms for approaches A and B

m	$\|b - A^2x_m^{new}\|_2$	$\|b - A^2x_m^{old}\|_2$
0	$0.21E2$	$0.21E2$
10	0.18	0.15
20	$0.27E-2$	$0.16E-1$
30	$0.53E-5$	$0.63E-2$
40	$0.16E-8$	$0.28E-2$
50		$0.36E-2$
60		$0.10E-2$
70		$0.49E-4$
80		$0.18e-5$
100		$0.21e-8$

(B) Generate the Krylov subspace for the operator A^2 and the vector b (the 'classical' approach). This leads to approximations denoted as x_m^{old}.

In Table 11.1 we have listed the norms of the residuals for the two approaches for some values of m. The analysis in [197] shows that the much faster convergence for the new approach could have been expected. Note that the new approach also has the advantage that there are only m matrix vector products. For the classical approach we need $2m$ matrix vector products with A, assuming that vectors like A^2p are computed by applying A twice. Usually, the matrix vector product is the CPU-dominating factor in the computations, since they operate in $\mathbb{R}^n$. The operations with $H_{m,m}$ are carried out in $\mathbb{R}^m$, and in typical applications $m \ll n$.

In [197] an example is also given for a more complicated function of A, namely the solution of

$$e^A x = b,$$

with A the same diagonal matrix as in the previous example, and b again the vector with all ones. This is a type of problem that is encountered in the solution of linear systems of ODEs. With the Krylov subspace for A and r_0 of dimension 20, a residual

$$\|r_m\|_2 \equiv \|b - e^A x_m\|_2 \approx 8.7E - 12$$

was observed, working in 48 bits floating point precision.

Others have also suggested working with the reduced system for the computation of, for instance, the exponential function of a matrix, as part of solution schemes for (parabolic) systems of equations. See, e.g., [117, 92, 137].

11.5 Matrix sign function

The matrix sign function $sign(A)$ for a nonsingular matrix A, with no eigenvalues on the imaginary axis, is defined as follows [15, 162]. Let

$$A = X\,\mathrm{diag}(J_+, J_-)X^{-1}$$

denote the decomposition of $A \in \mathbb{C}^{n\times n}$. The eigenvalues of J_+ lie in the right-half plane, and those of J_- are in the left-half plane. Let I_+ denote the identity matrix with the same dimensions as J_+, and I_- the identity matrix corresponding to J_-. Then

$$\mathrm{sign}(A) \equiv X\,\mathrm{diag}(I_+, -I_-)X^{-1}.$$

The sign function can be used, amongst others, to compute invariant subspaces, for instance those corresponding to the eigenvalues of A with positive real parts. It also plays an important role in QCD (cf. [147]). The Jordan decomposition of A is not a useful vehicle for the computation of this function. It can be shown that $\mathrm{sign}(A)$ is the limit of the Newton iteration

$$A_{k+1} = \tfrac{1}{2}(A_k + A_k^{-1}), \quad \text{for} \quad k = 0, 1, \ldots, \quad \text{with} \quad A_0 = A$$

see [15]. Unfortunately, the Newton iteration is also not suitable for accurate computation if A is ill-conditioned. Bai and Demmel consider more accurate ways of computation, which rely on the (block) Schur form of A:

$$B = Q^H A Q = \begin{bmatrix} B_{11} & B_{12} \\ 0 & B_{22} \end{bmatrix}.$$

The matrix Q is orthonormal, and it can easily be shown that

$$\mathrm{sign}(A) = Q\,\mathrm{sign}(B)Q^H.$$

Let this decomposition be such that B_{11} contains the eigenvalues of A with positive real part. Then let R be the solution of the Sylvester equation

$$B_{11}R - RB_{22} = -B_{12}.$$

This Sylvester equation can also be solved by a Newton iteration process. Then Bai and Demmel proved that

$$\mathrm{sign}(A) = Q \begin{bmatrix} I & -2R \\ 0 & -I \end{bmatrix} Q^H.$$

See [15] for further details, stability analysis, and examples of actual computations.

Suppose that we want to solve $\text{sign}(A)x = b$. Then, in view of the previous section, I suggest we start by constructing the Krylov subspace $K^m(A; b)$, and then compute the sign function for the reduced matrix $H_{m,m}$. This leads to the following approximation for the solution of $\text{sign}(A) = b$:

$$x_m = \|b\|_2 V_m \text{sign}(H_{m,m})^{-1} e_1. \tag{11.6}$$

Our preliminary experiments with this approach have been encouraging, for more details see [190].

12

Miscellaneous

12.1 Termination criteria

An important point, when using iterative processes, is to decide when to terminate the process. Popular stopping criteria are based on the norm of the current residual, or on the norm of the update to the current approximation to the solution (or a combination of these norms). More sophisticated criteria have been discussed in the literature.

In [124] a practical termination criterion for the conjugate gradient method is considered. Suppose we want an approximation x_i for the solution x for which

$$\|x_i - x\|_2 / \|x\|_2 \leq \varepsilon,$$

where ε is a tolerance set by the user.

It is shown in [124] that such an approximation is obtained by CG as soon as

$$\|r_i\|_2 \leq \mu_1 \|x_i\|_2 \varepsilon / (1 + \varepsilon),$$

where μ_1 stands for the smallest eigenvalue of the positive definite symmetric (preconditioned) matrix A. Of course, in most applications the value for μ_1 will be unknown, but with the iteration coefficients of CG we can build the tridiagonal matrix T_i, and compute the smallest eigenvalue (Ritz value) $\mu_1^{(i)}$ of T_i, which is an approximation for μ_1. In [124] a simple algorithm for the computation of $\mu_1^{(i)}$, along with the CG algorithm, is described, and it is shown that a rather robust stopping criterion is formed by

$$\|r_i\|_2 \leq \mu_1^{(i)} \|x_i\|_2 \varepsilon / (1 + \varepsilon).$$

A similar criterion had also been suggested earlier in [113].

It is tempting to use the same approach for the subspace methods for unsymmetric systems, but note that for such systems the norm of A^{-1} is not necessarily equal to the absolute value of some eigenvalue. Instead, we need the value of $\sqrt{\lambda_{max}(A^{-T}A^{-1})}$. It is tempting to approximate this value from spectral information of some reduced information. We could, for instance, use the eigenvalues of $H^T_{m+1,m}H_{m+1,m}$, from the upper Hessenberg matrix $H_{m+1,m}$ generated by one cycle of the GMRES(m) process, as approximations for eigenvalues of A^TA. This can be done, because the Arnoldi relation (3.18):

$$AV_m = V_{m+1}H_{m+1,m}$$

implies

$$V_m^TA^T = H^T_{m+1,m}V_{m+1},$$

and combining these two relations gives

$$V_mA^TAV_m = H^T_{m+1,m}H_{m+1,m}.$$

However, note that V_i has been generated for the Krylov subspace associated with A and v_1 and it is not clear how effective this basis is for the projection of A^TA. We have to build our own experiences with this for relevant classes of problems.

A quite different, but much more generally applicable, approach has been suggested in [5]. In this approach the approximate solution of an iterative process is regarded as the exact solution of some (nearby) linear system, and computable bounds for the perturbations with respect to the given system are presented. A nice overview of termination criteria has been presented in [20, Section 4.2].

12.2 Implementation aspects

For effective use of the given iteration schemes, it is necessary that they can be implemented in such a way that high computing speeds are achievable. It is most likely that high computing speeds will be realized only by parallel architectures and therefore we must see how well iterative methods fit to such computers.

The iterative methods only need a handful of basic operations per iteration step

(1) Vector updates: In each iteration step the current approximation to the solution is updated by a correction vector. Often the corresponding residual vector is also obtained by a simple update, and we also have update formulas for the correction vector (or search direction).

(2) Inner products: In many methods the speed of convergence is influenced by carefully constructed iteration coefficients. These coefficients are sometimes known analytically, but more often they are computed by inner products, involving residual vectors and search directions, as in the methods discussed in the previous sections.
(3) Matrix vector products: In each step at least one matrix vector product has to be computed with the matrix of the given linear system. Sometimes matrix vector products with the transpose of the given matrix are also required (e.g., Bi-CG). Note that it is not necessary to have the matrix explicitly, it suffices to be able to generate the result of the matrix vector product.
(4) Preconditioning: It is common practice to precondition the given linear system by some preconditioning operator. Again it is not necessary to have this operator in explicit form, it is enough to generate the result of the operator applied to some given vector. The preconditioner is applied as often as the matrix vector multiply in each iteration step.

For large enough problem sizes the inner products, vector updates, and matrix vector product are easily parallelized and vectorized. The more successful preconditionings, i.e., based upon incomplete LU decomposition, are not easily parallelizable. For that reason often the use of only diagonal scaling as a preconditioner on highly parallel computers, such as the CM2, is satisfactory [27].

On distributed memory computers we need large grained parallelism in order to reduce synchronization overheads. This can be achieved by combining the work required for a successive number of iteration steps. The idea is first to construct in parallel a straightforward Krylov basis for the search subspace in which an update for the current solution will be determined. Once this basis has been computed, the vectors are orthogonalized, as is done in Krylov subspace methods. The construction, as well as the orthogonalization, can be done with large grained parallelism, and contains a sufficient degree of parallelism.

This approach has been suggested for CG in [40] and for GMRES in [41], [16], and [52]. One of the disadvantages of this approach is that a straightforward basis, of the form $y, Ay, A^2y, \ldots, A^iy$, is usually very ill-conditioned. This is in sharp contrast to the optimal condition of the orthogonal basis set constructed by most of the projection type methods and it puts severe limits on the number of steps that can be combined. However, in [16] and [52], ways of improving the condition of a parallel generated basis are suggested and it seems possible to take larger numbers of steps, say 25, together. In [52], the effects of this approach on the communication overheads are studied and compared with experiments done on moderately massive parallel transputer systems.

12.3 Parallelism and data locality in CG

For successful application of CG the matrix A must be symmetric positive definite. In other short recurrence methods, other properties of A may be desirable, but we will not exploit these properties explicitly in the discussion on parallel aspects.

Most often, the conjugate gradients method is used in combination with some kind of preconditioning. This means that the matrix A can be thought of as being multiplied by some suitable approximation K^{-1} for A^{-1}. Usually, K is constructed as an approximation of A, such that systems like $Ky = z$ are much easier to solve than $Ax = b$. Unfortunately, a popular class of preconditioners, based upon an incomplete factorization of A, do not lend themselves well to parallel implementation.

For the moment we will assume that the preconditioner is chosen such that the parallelism in solving $Ky = z$ is comparable with the parallelism in computing Ap, for given p.

For CG it is also required that the preconditioner K be symmetric positive definite. This aspect will play a role in our discussions because it shows how some properties of the preconditioner can sometimes be used to our advantage for an efficient implementation.

The scheme for preconditioned CG is given in Figure 5.2. Note that in that scheme the updating of x and r can only start after the completion of the inner product required for α_i. Therefore, this inner product is a so-called synchronization point: all computation has to wait for completion of this operation. We can, as much as possible, try to avoid such synchronization points, or to formulate CG in such a way that synchronization points can be taken together. We will look at such approaches later.

Since on a distributed memory machine communication is required to assemble the inner product, it would be nice if we could proceed with other useful computation while the communication takes place. However, as we see from our CG scheme, there is no possibility of overlapping this communication time with useful computation. The same observation can be made for the updating of p, which can only take place after the completion of the inner product for β_i. Apart from the computation of Ap and the computations with K, we need

x_0 is an initial guess; $r_0 = b - Ax_0$;
$q_{-1} = p_{-1} = 0$; $\beta_{-1} = 0$;
Solve w_0 from $Kw_0 = r_0$;
$s_0 = Aw_0$;
$\rho_0 = (r_0, w_0)$; $\mu_0 = (s_0, w_0)$;
$\alpha_0 = \rho_0/\mu_0$;
for $i = 0, 1, 2, \ldots.$
  $p_i = w_i + \beta_{i-1} p_{i-1}$;
  $q_i = s_i + \beta_{i-1} q_{i-1}$;
  $x_{i+1} = x_i + \alpha_i p_i$;
  $r_{i+1} = r_i - \alpha_i q_i$;
  if x_{i+1} accurate enough **then** quit;
  Solve w_{i+1} from $Kw_{i+1} = r_{i+1}$;
  $s_{i+1} = Aw_{i+1}$;
  $\rho_{i+1} = (r_{i+1}, w_{i+1})$;
  $\mu_{i+1} = (s_{i+1}, w_{i+1})$;
  $\beta_i = \frac{\rho_{i+1}}{\rho_i}$;
  $\alpha_{i+1} = \frac{\rho_{i+1}}{\mu_{i+1} - \rho_{i+1}\beta_i/\alpha_i}$;
end i;

Figure 12.1. Parallel CG; Chronopoulos and Gear variant.

to load seven vectors for ten vector floating point operations. This means that for this part of the computation only $\frac{10}{7}$ floating point operations can be carried out per memory reference on average.

Several authors [40, 141, 142] have attempted to improve this ratio, and to reduce the number of synchronization points. In our formulation of CG there are two such synchronization points, namely the computation of both inner products. Meurant [141] (see also [163]) has proposed a variant in which there is only one synchronization point, though at the cost of possibly reduced numerical stability, and one additional inner product. In this scheme the ratio between computations and memory references is about 2.

In Figure 12.1, we show a variant that was proposed by Chronopoulos and Gear [40]. In this scheme all vectors need be loaded only once per pass of the loop, which leads to better exploitation of the data (improved data locality). However, the price is that we need another $2n$ flops per iteration step. Chronopoulos and Gear [40] claim stability, based upon their numerical experiments.

Instead of two synchronization points, as in the standard version of CG, we now have only one synchronization point, as the next loop can only be started when the inner products at the end of the previous loop have been assembled. Another slight advantage is that these inner products can be computed in parallel.

Chronopoulos and Gear [40] propose to improve further the data locality and parallelism in CG by combining s successive steps. Their algorithm is based upon the following property of CG. The residual vectors $r_0, \ldots, r_i$ form an orthogonal basis (assuming exact arithmetic) for the Krylov subspace spanned by $r_0, Ar_0, \ldots, A^{i-1}r_0$. When r_j is reached, the vectors $r_0, r_1, \ldots, r_j, Ar_j, \ldots, A^{i-j-1}r_j$ also form a basis for this subspace. Hence, we may combine s successive steps by first generating $r_j, Ar_j, \ldots, A^{s-1}r_j$, and then doing the orthogonalization and the updating of the current solution with this blockwise extended subspace. This approach leads to a slight increase in flops in comparison with s successive steps of the standard CG, and also one additional matrix vector product is required per s steps.

The main drawback in this approach seems to be potential numerical instability. Depending on the spectral properties of A, the set $r_j, \ldots, A^{s-1}r_j$ may tend to converge to a vector in the direction of a dominating eigenvector, or, in other words, may tend to dependence for increasing values of s. The authors claim to have seen successful completion of this approach, with no serious stability problems, for small values of s. Nevertheless, it seems that s-step CG, because of these problems, has a bad reputation (see also [164]). However, a similar approach, suggested by Chronopoulos and Kim [41] for other processes such as GMRES, seems to be more promising. Several authors have pursued this research direction, and we will come back to this in Section 12.5.

We consider yet another variant of CG, in which there is a possibility of overlapping all of the communication time with useful computations [56]. This variant, represented in Figure 12.2, is just a reorganized version of the original CG scheme, and is therefore roughly as stable. The key trick in this approach is to delay the updating of the solution vector by one iteration step. In fact, it is somewhat more stable because the inner product for ρ is computed in a more stable way. The computation guarantees that the inner product is always positive, even for highly ill-conditioned K.

Another advantage over the previous scheme is that no additional operations are required. It is assumed that the preconditioner K can be written as $K = LL^T$. Furthermore, it is assumed that the preconditioner has a block structure, corresponding to the gridblocks assigned to the processors, so that communication (if necessary) can be overlapped with computation.

x_0 is an initial guess; $r_0 = b - Ax_0$;
$p_{-1} = 0$; $\beta_{-1} = 0$; $\alpha_{-1} = 0$;
$s = L^{-1} r_0$;
$\rho_0 = (s, s)$;
for $i = 0, 1, 2, ...$

$w_i = L^{-T} s$;	(0)
$p_i = w_i + \beta_{i-1} p_{i-1}$;	(1)
$q_i = A p_i$;	(2)
$\gamma = (p_i, q_i)$;	(3)
$x_i = x_{i-1} + \alpha_{i-1} p_{i-1}$;	(4)
$\alpha_i = \frac{\rho_i}{\gamma}$;	(5)
$r_{i+1} = r_i - \alpha_i q_i$;	(6)
$s = L^{-1} r_{i+1}$;	(7)
$\rho_{i+1} = (s, s)$;	(8)
if r_{i+1} *small enough* **then**	(9)

$x_{i+1} = x_i + \alpha_i p_i$
quit;
$\beta_i = \frac{\rho_{i+1}}{\rho_i}$;
end i;

Figure 12.2. Parallel CG; Demmel et al. variant.

Now I discuss how this scheme may lead to an efficient parallel scheme, and how local memory (vector registers, cache, ...) can be exploited.

(1) All computing intensive operations can be carried out in parallel. Communication between processors is only required for the operations (2), (3), (7), (8), (9), and (0). I have assumed that the communication in (2), (7), and (0) can be largely overlapped with computation.
(2) The communication required for the assembly of the inner product in (3) can be overlapped with the update for x (which could have been done in the previous iteration step).
(3) The assembly of the inner product in (8) can be overlapped with the computation in (0). Also step (9) usually requires information such as the norm of the residual, which can be overlapped with (0).
(4) Steps (1), (2), and (3) can be combined: the computation of a segment of p_i can be followed immediately by the computation of a segment of q_i in (2), and this can be followed by the computation of a part of the inner product in (3). This saves on load operations for segments of p_i and q_i.

(5) Depending on the structure of L, the computation of segments of r_{i+1} in (6) can be followed by operations in (7), which can be followed by the computation of parts of the inner product in (8), and the computation of the norm of r_{i+1}, required for (9).

(6) The computation of β_i can be done as soon as the computation in (8) has been completed. At that moment, the computation for (1) can be started if the requested parts of w_i have been completed in (0).

(7) If no preconditioner is used, then $w_i = r_i$, and steps (7) and (0) have to be skipped. Step (8) has to be replaced by $\rho_i = (r_{i+1}, r_{i+1})$. Now we need useful computation in order to overlap the communication for this inner product. To this end, we might split the computation in (4) per processor into two parts. The first of these parts is computed in parallel in overlap with (3), while the parallel computation of the other parts is used in order to overlap the communication for the computation of ρ_i.

12.4 Parallel performance of CG

Some realistic 3D computational fluid dynamics simulation problems, as well as other problems, lead to the necessity to solve linear systems $Ax = b$ with a matrix of very large order, say, billions of unknowns. If not of very special structure, such systems are not likely to be solved by direct elimination methods. For such very large (sparse) systems we must exploit parallelism in combination with suitable solution techniques, such as iterative solution methods.

From a parallel point of view CG mimics very well parallel performance properties of a variety of iterative methods such as Bi-CG, CGS, Bi-CGSTAB, QMR, and others.

In this section I study the performance of CG on parallel distributed memory systems and I report on some supporting experiments on actual existing machines. Guided by experiments I will discuss the suitability of CG for massively parallel processing systems.

All computationally intensive elements in preconditioned CG (updates, inner products, and matrix vector operations) are trivially parallelizable for shared memory machines, see [60], except possibly for the preconditioning step: *Solve* w_{i+1} *from* $Kw_{i+1} = r_{i+1}$. For the latter operation parallelism depends very much on the choice for K. In this section we restrict ourselves to block Jacobi preconditioning, where the blocks have been chosen so that each processor can handle one block independently of the others. For other preconditioners that allow some degree of parallelism, see [60].

For a distributed memory machine at least some of the steps require communication between processors: the accumulation of inner products and the computation of Ap_i (depending on the nonzero structure of A and the distribution of the nonzero elements over the processors). We consider in more detail the situation where A is a block-tridiagonal matrix of order N, and we assume that all blocks are of order $\sqrt{N}$:

$$A = \begin{pmatrix} A_1 & D_1 & & \\ D_1 & A_2 & D_2 & \\ & D_2 & \ddots & \ddots \\ & & \ddots & \\ & & & \end{pmatrix},$$

in which the D_i are diagonal matrices, and the A_i are tridiagonal matrices. Such systems occur quite frequently in finite difference approximations in 2 space dimensions. Our discussion can easily be adapted to 3 space dimensions.

12.4.1 Processor configuration and data distribution

For simplicity I will assume that the processors are connected as a 2D grid with $p \times p = P$ processors.

The data have been distributed in a straightforward manner over the processor memories and I have not attempted to exploit fully the underlying grid structure for the given type of matrix (in order to reduce communication as much as possible). In fact it will turn out that in our case the communication for the matrix vector product plays only a minor role for matrix systems of large size.

Because of symmetry only the 3 nonzero diagonals in the upper triangular part of A need to be stored, and we have chosen to store successive parts of length N/P of each diagonal in consecutive neighbouring processors. In Figure 12.3 we see which part of A is represented by the data in the memory of a given processor.

The blocks for block Jacobi are chosen to be the diagonal blocks that are available on each processor, and the various vectors (r_i, p_i, etc.), have been distributed likewise, i.e., each processor holds a section of length N/P of these vectors in its local memory.

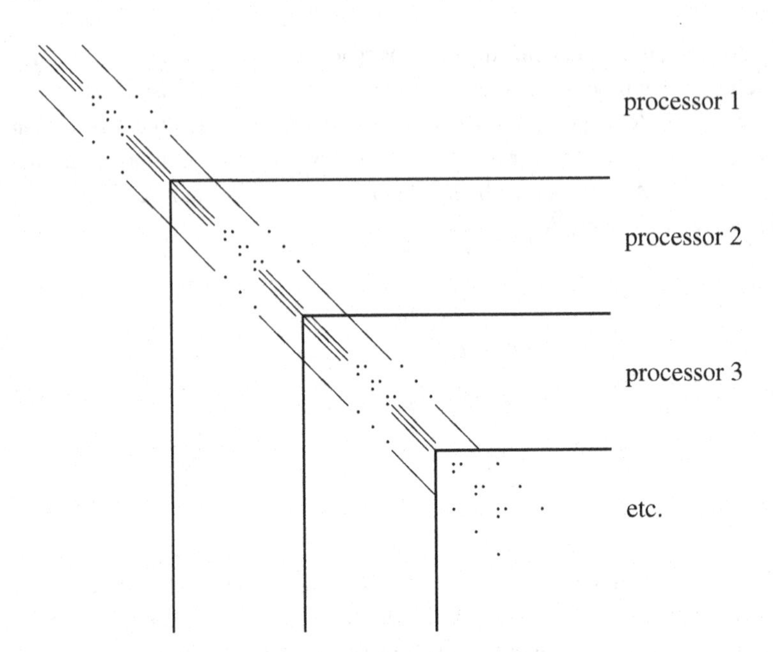

Figure 12.3. Distribution of A over the processors.

12.4.2 Required communication

Matrix vector product It is easily seen for a $2D$ processor grid (as well as for many other configurations, including hypercube and pipeline), that the matrix vector product can be completed with only neighbour–neighbour communication. This means that the communication costs do not increase for increasing values of p. If we follow a domain decomposition approach, in which the finite difference discretization grid is subdivided into p by p subgrids (p in x-direction and p in y-direction), then the communication costs are smaller than the computational costs by a factor of $\mathcal{O}(\frac{\sqrt{N}}{p})$.

In [51] much attention is given to this sparse matrix vector product and it is shown that the time for communication can be almost completely overlapped with computational work. Therefore, with adequate coding the matrix vector products do not necessarily lead to serious communication problems, not even for relatively small-sized problems.

Vector update In our case these operations do not require any communication and we should expect linear speedup when increasing the number of processors P.

Inner product For the inner product we need global communication for the assembly and we need global communication for the distribution of the assembled inner product over the processors. For a $p \times p$ processor grid these communication costs are proportional to p. This means that for a constant length of the vector parts per processor, these communication costs will dominate for large enough values of p. This is quite unlike the situation for the matrix vector product and as we will see it may be a severely limiting factor in achieving high speedups in a massively parallel environment.

The experiments, reported in [47, 48], and the modelling approach (see also [55]) clearly show that even a method like CG, which might be anticipated to be highly parallel, may suffer severely from the communication overhead due to the required inner products. These studies indicate that if we want reasonable speedup in a massively parallel environment then the local memories should also be much larger when the number of processors is increased in order to accommodate systems large enough to compensate for the increased global communication costs.

Another approach for the reduction of the relative costs for communication is to do more useful computational work per iteration step, so that the communication for the two inner products takes relatively less time. One way to do this is to use polynomial preconditioning, i.e., the preconditioner consists of a number of matrix vector products with the matrix A. This may work well in situations where the matrix vector product requires only little (local) communication. Another way is to apply domain decomposition: the given domain is split into P, say, subdomains with estimated values for the solutions on the interfaces. Then all the subproblems are solved independently and in parallel. This way of approximating the solution may be viewed as a preconditioning step in an iterative method. In this way we do more computational work per communication step. Unfortunately, depending on the problem and on the way of decoupling the subdomains, we may need a larger number of iteration steps for larger values of P, which may then, of course, diminish the overall efficiency of the domain decomposition approach. For more information on this approach see references given in [20].

12.5 Parallel implementation of GMRES(m)

The time consuming part of GMRES(m) is the construction of a basis for the Krylov subspace. We isolate this for one full cycle in Figure 12.4.

Obviously, we need $\frac{1}{2}(m^2 + 3m)$ inner products and vector updates per cycle, and $m + 1$ matrix vector operations (and possibly $m + 1$ operations with a

$r = b - Ax_0$, for a given initial guess x_0
$v_1 = r/\|r\|_2$
for $i = 1, 2, ..., m$
  $\widehat{v}_{i+1} = Av_i$
  for $k = 1, ..., i$
    $h_{k,i} = v_k^T \widehat{v}_{i+1}$
    $\widehat{v}_{i+1} = \widehat{v}_{i+1} - h_{k,i} v_k$
  end
  $h_{i+1,i} = ||\widehat{v}_{i+1}||_2$
  $v_{i+1} = \widehat{v}_{i+1}/h_{i+1,i}$
end

Figure 12.4. One cycle of GMRES(m) with modified Gram–Schmidt.

preconditioner). Much can be done in parallel, depending on the structure of the matrix and the preconditioner, but a bottleneck with respect to communication and coarse grain parallelism is the inner **for** loop. The inner products and the vector updates have to be done consecutively.

We now follow an approach for more coarse grained parallelism that has been proposed in [55]. The inner loop is modified Gram–Schmidt for the basis of the Krylov subspace and stable computation requires that the expansion of the subspace is done with the last computed orthogonal vector v_i. If a set of nonorthogonal basis vectors $\widehat{v}_1, \widehat{v}_2, \ldots, \widehat{v}_{m+1}$ is available at the start of a cycle, then the modified Gram–Schmidt process can be rearranged as in Figure 12.5.

Exercise 12.1. *Verify that the Modified Gram–Schmidt process in Figure 12.5 leads to the same orthogonal basis, even in rounded arithmetic, as the process in Figure 12.4, if $\widehat{v}_{i+1}$ is given as the same vector at the start of Figure 12.5 (instead of being computed as Av_i).*

Note that the inner products in the **for** k loop in Figure 12.5 can be executed in parallel. This means that their values can be combined in one single communication step at the end of the loop, so that communication can be reduced to m messages instead of $\frac{1}{2}(m^2 + 3m)$ messages. The new rearranged Gram–Schmidt process allows also for overlap of communication with computations, see [55].

There are two problems with this parallelizable approach. The first one is that we need a well-conditioned set of basis vectors $\widehat{v}_j$ to start with. This can be handled in different ways. The worst possibility is to use the standard

$r = b - Ax_0$, for a given initial guess x_0
$v_1 = r/\|r\|_2$
for $i = 1, 2, \ldots, m$
    for $k = i + 1, \ldots, m + 1$
        $g_{i,k-1} = v_i^T \widehat{v}_k$
        $\widehat{v}_k = \widehat{v}_k - g_{i,k} v_i$
    end
    $g_{i+1,i} = \|\widehat{v}_{i+1}\|_2$
    $v_{i+1} = \widehat{v}_{i+1}/g_{i+1,i}$
end

Figure 12.5. Rearranged modified Gram–Schmidt.

Krylov vectors $\widehat{v}_j = A^{j-1}\widehat{v}_1$. In [16] it is suggested that a reasonably well-conditioned basis with spectral information of A is constructed. A good and very easy alternative is the following one. The first cycle of GMRES is done in the standard way (with less exploitable parallelism) and then in each new cycle the vector $\widehat{v}_j$ is generated by the **for** i loop in Figure 12.4 in which the computation of the inner products is skipped. For the $h_{i,j}$ elements simply take the elements of the previous GMRES(m) cycle. A similar approach has been advocated for the construction of the GMRES(ℓ) part in Bi-CGSTAB(ℓ) [177].

The other problem is that the inner products in the **for** k loop in Figure 12.5 do not represent the required upper Hessenberg matrix that represents the projection of A with respect to the final orthogonal basis. In [55, Section 4.4] the formulas for the construction of this upper Hessenberg matrix are described. It is shown, by numerical experiments carried out on a 400-processor Parsytec Supercluster, that the restructured process may be significantly more efficient than the classical process. Gains in speed by a factor of about 2 are observed for $n = 10\,000$ and $m = 30, 50$ in GMRES(m) [55].

13

Preconditioning

13.1 Introduction

As we have seen in our discussions on the various Krylov subspace methods, they are not robust in the sense that they can be guaranteed to lead to acceptable approximate solutions within modest computing time and storage (modest with respect to alternative solution methods). For some methods (for instance, full GMRES) it is obvious that they lead, in exact arithmetic, to the exact solution in maximal n iterations, but that may not be very practical. Other methods are restricted to specific classes of problems (CG, MINRES) or suffer from such nasty side-effects as stagnation or breakdown (Bi-CG, Bi-CGSTAB). Such poor convergence depends in a very complicated way on spectral properties (eigenvalue distribution, field of values, condition of the eigensystem, etc.) and this information is not available in practical situations.

The trick is then to try to find some nearby operator K such that $K^{-1}A$ has better (but still unknown) spectral properties. This is based on the observation that for $K = A$, we would have the ideal system $K^{-1}Ax = Ix = K^{-1}b$ and all subspace methods would deliver the true solution in one singe step. The hope is that for K in some sense close to A a properly selected Krylov method applied to, for instance, $K^{-1}Ax = K^{-1}b$, would need only a few iterations to yield a good enough approximation for the solution of the given system $Ax = b$. An operator that is used for this purpose is called a *preconditioner* for the matrix A.

The general problem of finding an efficient preconditioner, is to identify a linear operator K (the *preconditioner*) with the properties that[1]:

(1) K is a good approximation to A in some sense.

[1] The presentation in this chapter has partial overlap with [61, Chapter 9].

(2) The cost of the construction of K is not prohibitive.
(3) The system $Ky = z$ is much easier to solve than the original system.

Research on preconditioning is a very broad and active area of research with only little structure. There is no general theory on which we can safely base an efficient selection. The main difficulty is that preconditioning is based on approximation and in the absence of precise information on the behaviour of the solution of a given system $Ax = b$ and on the spectral properties of A, it may occur that the convergence depends critically on the information that is discarded in the approximation process. Selection and construction of a good preconditioner for a given class of problems is therefore at best an educated guess. It is not my aim to give a complete overview of all existing preconditioning techniques. Instead I will consider the main ideas in order to guide the reader in the construction or selection of a proper preconditioner.

There is a great freedom in the definition and construction of preconditioners for Krylov subspace methods and that is one reason why these methods are so popular and so successful. Note that in all the Krylov methods, we never need to know individual elements of A, and we never have to modify parts of the given matrix. It is always sufficient to have a rule (subroutine) that generates, for given input vector y, the output vector z that can *mathematically* be described as $z = Ay$. This also holds for the nearby operator: it does not have to be an explicitly given matrix. However, it should be realized that the operator (or subroutine) that generates the approximation for A can be mathematically represented as a matrix. It is then important to verify that application of the operator (or subroutine, or possibly even a complete code) on different inputs leads to outputs that have the same mathematical relation through some (possibly explicitly unknown) matrix K. For some methods, in particular Flexible GMRES and GMRESR, it is permitted that the operator K is (slightly) different for different input vectors (*variable preconditioning*). This plays an important role in the solution of nonlinear systems, if the Jacobian of the system is approximated by a Frechet derivative and it is also attractive in some domain decomposition approaches (in particular, if the solution per domain itself is again obtained by some iterative method).

The following aspect is also important. Except for some trivial situations, the matrix $K^{-1}A$ is never formed explicitly. In many cases this would lead to a dense matrix and destroy all efficiency that could be obtained for the often sparse A. Even for dense matrix A it might be too expensive to form the preconditioned matrix explicitly. Instead, for each required application of $K^{-1}A$ to some vector y, we first compute the result w of the operator A applied to y and then we determine the result z of the operator K^{-1} applied to w. This is often done by solving z from $Kz = w$, but there are also approaches by which approximations M for A^{-1} are constructed (e.g., *sparse approximate inverses*) and then we apply, of course, the operator M to w in order to obtain z. Only very special and simple preconditioners like diagonal matrices can be applied explicitly to A. This can be done before and in addition to the construction of another preconditioning.

Remember always that whatever preconditioner we construct, the goal is to reduce CPU time (or memory storage) for the computation of the desired approximated solution.

There are different ways of implementing preconditioning; for the same preconditioner these different implementations lead to the same eigenvalues for the preconditioned matrices. However, the convergence behaviour is also dependent on the eigenvectors or, more specifically, on the components of the starting residual in eigenvector directions. Since the different implementations can have quite different eigenvectors, we may thus expect that their convergence behaviour can be quite different. Three different implementations are as follows:

(1) **Left-preconditioning**: Apply the iterative method to $K^{-1}Ax = K^{-1}b$. We note that symmetry of A and K does not imply symmetry of $K^{-1}A$. However, if K is symmetric positive definite then $[x, y] \equiv (x, Ky)$ defines a proper inner product. It is easy to verify that $K^{-1}A$ is symmetric with respect to the new inner product $[\,,\,]$, so that we can use methods like MINRES, SYMMLQ, and CG (when A is also positive definite) in this case. Popular formulations of preconditioned CG are based on this observation, see Section 5.2.

    If we are using a minimal norm residual method (for instance GMRES or MINRES), we should note that with left-preconditioning we are minimizing the preconditioned residual $K^{-1}(b - Ax_k)$, which may be quite different from the residual $b - Ax_k$. This could have consequences for stopping criteria that are based on the norm of the residual.

(2) **Right-preconditioning**: Apply the iterative method to $AK^{-1}y = b$, with $x = K^{-1}y$. This form of preconditioning also does not lead to a symmetric product when A and K are symmetric.

    With right-preconditioning we have to be careful with stopping criteria that are based upon the error: $||y - y_k||_2$ may be much smaller than the error-norm $||x - x_k||_2$ (equal to $||K^{-1}(y - y_k)||_2$) that we are interested in. Right-preconditioning has the advantage that it only affects the operator and not the right-hand side. This may be an attractive property in the design of software for specific applications.

(3) **Two-sided preconditioning**: For a preconditioner K with $K = K_1 K_2$, the iterative method can be applied to $K_1^{-1} A K_2^{-1} z = K_1^{-1} b$, with $x = K_2^{-1} z$. This form of preconditioning may be used for preconditioners that come in factored form. It can be seen as a compromise between left- and right-preconditioning. This form may be useful for obtaining a (near) symmetric operator for situations where K cannot be used for the definition of an inner product (as described under left-preconditioning).

Note that with all these forms of preconditioning, either explicit or implicit, we are generating, through a redefinition of the inner product, a Krylov subspace for the preconditioned operator. This implies that the reduced matrix $H_{i,i}$ (cf. (3.18), gives information about the preconditioned matrix: in particular, the Ritz values approximate eigenvalues of the preconditioned matrix. The generated Krylov subspace cannot be used in order to obtain information as well for the unpreconditioned matrix.

The choice of K varies from purely 'black box' algebraic techniques that can be applied to general matrices to 'problem dependent' preconditioners that exploit special features of a particular problem class. Examples of the last class are discretized partial differential equations, where the preconditioner is constructed as the discretization of a nearby (easier to solve) PDE. Although problem dependent preconditioners can be very powerful, there is still a practical need for efficient preconditioning techniques for large classes of problems.

We will now discuss some seemingly strange effects of preconditioning. There is very little theory for what we can expect a priori with a specific type of preconditioner. It is well known that incomplete LU decompositions exist if the matrix A is an M-matrix, but that does not say anything about the potential reduction in the number of iterations. For the discretized Poisson equation, it has been proved [175] that the number of iterations will be reduced by a factor larger than 3. This leads to a true reduction in CPU time, because the complexity per (preconditioned) iteration increases with a factor of about 2.

For systems that are not positive definite, almost anything can happen. For instance, let us consider a symmetric matrix A that is indefinite. The goal of preconditioning is to approximate A by K, and a common strategy is to ensure that the preconditioned matrix $K^{-1}A$ has its eigenvalues clustered near 1 as much as possible. Now imagine some preconditioning process in which we can improve the preconditioner continuously from $K = I$ to $K = A$. For instance, we might think of incomplete LU factorization with a drop-tolerance criterion. For $K = I$, the eigenvalues of the preconditioned matrix are clearly those of A and thus are at both sides of the origin. Since eventually when the preconditioner is equal to A all eigenvalues are exactly 1, the eigenvalues have to move gradually in the direction of 1, as the preconditioner is improved. The negative eigenvalues, on their way towards 1 have to pass the origin, which means that while improving the preconditioner the preconditioned matrix may from time to time have eigenvalues very close to the origin. In my chapter on iterative methods, I have explained that the residual in the i-th iteration step can be expressed as

$$r_i = P_i(B)r_0,$$

where B represents the preconditioned matrix. Since the polynomial P_i has to satisfy $P_i(0) = 1$, and since the values of P_i should be small on the eigenvalues of B, this may help to explain why there may not be much reduction for components in eigenvector directions corresponding to eigenvalues close to zero, if i is still small. This means that, when we improve the preconditioner, in the sense that the eigenvalues are getting more clustered towards 1, its effect on the iterative method may be dramatically worse for some 'improvements'. This is a qualitative explanation of what we have observed many times in practice. By increasing the number of fill-in entries in ILU, sometimes the number of iterations increases. In short, the number of iterations may be a very irregular function of the level of the incomplete preconditioner. For other types of preconditioners similar observations may be made.

There are only very few specialized cases where it is known a priori how to construct a good preconditioner and there are few proofs of convergence except in very idealized cases. For a general system, however, the following approach may help to build up our insight into what is happening. For a representative linear system, we start with unpreconditioned GMRES(m), with m as high as possible. In one cycle of GMRES(m), the method explicitly constructs an upper Hessenberg matrix of order m, denoted by H_m. This matrix is reduced to upper triangular form but, before this takes place, we should compute the eigenvalues of H_m, called the Ritz values. These Ritz values usually give a fairly good impression of the most relevant parts of the spectrum of A. Then we do the same with the preconditioned system and inspect the effect on the spectrum. If there is no specific trend of improvement in the behaviour of the Ritz values, when we try to improve the preconditioner, then obviously we have to look for another class of preconditioner. If there is a positive effect on the Ritz values, then this may give us some insight as to by how much the preconditioner has to be further improved in order to be effective. At all times, we have to keep in mind the rough analysis that we made in this chapter, and check whether the construction of the preconditioner and its costs per iteration are still inexpensive enough to be amortized by an appropriate reduction in the number of iterations.

In this chapter I will describe some of the more popular preconditioning techniques and give references and pointers for other techniques. I refer the reader to [10, 37, 168, 144] for more complete overviews of (classes of) preconditioners. See [22] for a very readable introduction to various concepts of preconditioning and for many references to specialized literature.

Originally, preconditioners were based on direct solution methods in which part of the computation is skipped. This leads to the notion of *Incomplete LU* (or *ILU*) *factorization* [139, 10, 168]. I will now discuss these incomplete factorizations in more detail.

13.2 Incomplete LU factorizations

Standard Gaussian elimination is equivalent to factoring the matrix A as $A = LU$, where L is lower triangular and U is upper triangular. In actual computations these factors are explicitly constructed. The main problem in sparse matrix computations is that the factors of A are often a good deal less sparse than A, which makes solution expensive. The basic idea in the point ILU preconditioner is to modify Gaussian elimination to allow fill-ins at only a restricted set of positions in the LU factors. Let the allowable fill-in positions be given by the index set S, i.e.

$$\begin{aligned} l_{i,j} &= 0 \quad \text{if} \quad j > i \quad \text{or} \quad (i, j) \notin S \\ u_{i,j} &= 0 \quad \text{if} \quad i > j \quad \text{or} \quad (i, j) \notin S. \end{aligned} \tag{13.1}$$

A commonly-used strategy is to define S by:

$$S = \{(i, j) | \quad a_{i,j} \neq 0\}. \tag{13.2}$$

That is, the only nonzeros allowed in the LU factors are those for which the corresponding entries in A are nonzero. Before we proceed with different strategies for the construction of effective incomplete factorizations we consider whether these factorizations exist. It can be shown that they exist for so-called M-matrices.

The theory of M-matrices for iterative methods is very well covered by Varga [212]. These matrices occur frequently after discretization of partial differential equations, and for M-matrices we can identify all sorts of approximating matrices K for which the basic splitting leads to a convergent iteration (3.1). For preconditioners for Krylov subspace methods it is not important that the basic iteration converges; primarily we want reduced condition numbers and/or better eigenvalue distributions for the preconditioned matrices. These latter properties are very difficult to prove. In fact, some of these effects have been proven only for very special model problems, for examples of this, see [195, 205].

I will now first show how the M-matrix theory functions in the construction of incomplete LU factorizations. I consider first the complete Gaussian elimination process, for which the following extension of a theorem by Fan [80, p.44] is useful [139].

Theorem 13.1. *Gaussian elimination preserves the M-matrix property.*

Proof. We consider one step in the elimination process: the elimination of the subdiagonal elements in the first column of A. In matrix notation this leads to

$$A^{(1)} = L^{(1)}, \tag{13.3}$$

with

$$L^{(1)} = \begin{pmatrix} 1 & & & \\ -\frac{a_{2,1}}{a_{1,1}} & 1 & & \\ \vdots & & \ddots & \\ -\frac{a_{n,1}}{a_{1,1}} & & & 1 \end{pmatrix} \geq 0\ .$$

The elements of $A^{(1)}$ are given by $a_{i,j}^{(1)} = a_{i,j} - \frac{a_{1,j}}{a_{1,1}} a_{i,1}$, for $i > 1$, and $a_{1,j}^{(1)} = a_{1,j}$.

Note that $a_{i,j}^{(1)} \leq 0$ for $i \neq j$.

It remains to be shown that $(A^{(1)})^{-1} \geq 0$. We consider the i-th column

$$(A^{(1)})^{-1} e_i = A^{-1} (L^{(1)})^{-1} e_i.$$

For $i = 1$ it follows that $A^{-1}(L^{(1)})^{-1} e_1 = \frac{1}{a_{1,1}} e_1$. For $i \neq 1$ we have that $(A^{(1)})^{-1} e_i = A^{-1} e_i \geq 0$. □

After an elimination step we ignore certain fill-in elements in off-diagonal positions. Now the following (almost trivial) extension of a theorem of Varga [212, Theorem 3.12] is helpful [139, Theorem 2.2].

Theorem 13.2. *If $A = (a_{i,j})$ is a real $n \times n$ M-matrix and $C = (c_{i,j})$ is a real $n \times n$ matrix with $c_{i,i} \geq a_{i,i}$ and $a_{i,j} \leq c_{i,j} \leq 0$, for $i \neq j$, then C is also an M-matrix.*

Exercise 13.1. *Suppose that we replace an off-diagonal element in the second row of $A^{(1)}$ by zero. Let the result after the elimination of the second column of* *the remaining matrix be denoted by $\widetilde{A}^{(2)}$. Show that*

$$\widetilde{a}_{i,i}^{(2)} \geq a_{i,i}^{(2)}, \tag{13.4}$$

for all i, and

$$a_{i,j}^{(2)} \leq \widetilde{a}_{i,j}^{(2)} \leq 0, \tag{13.5}$$

for $i \neq j$.

Theorem 13.1 says that after a Gaussian elimination step on an M-matrix A, we obtain a reduced matrix that is still an M-matrix. The relations (13.4) and (13.5), together with Theorem 13.2, show that even after ignoring off-diagonal elements in the reduced matrix we still have an M-matrix.

Exercise 13.2. *Suppose that A is an M-matrix. Consider the Gaussian elimination corrections to the diagonal of A after one step of Gaussian elimination (that is, consider the diagonal elements of $A^{(1)} - A$). Show that after ignoring some of these diagonal corrections the remaining matrix has still the M-matrix property.*

After $n-1$ steps of Gaussian elimination we obtain an upper triangular matrix

$$U \equiv A^{(n-1)} = L^{(n-1)}L^{(n-2)} \cdots L^{(1)}A, \tag{13.6}$$

with

$$L^{-1} \equiv L^{(n-1)}L^{(n-2)} \cdots L^{(1)}. \tag{13.7}$$

This gives the decomposition $A = LU$. Repeated application of Theorem 13.1 leads to the observation that U is an M-matrix. Because of the above given arguments, we conclude that U is still an M-matrix after ignoring Gaussian elimination corrections in various stages of the elimination process.

Now we consider the elimination matrices $L^{(i)}$. Obviously the elements of these matrices are nonnegative.

The following exercise involves the proof of a very well-known fact (see, for example, [98]).

Exercise 13.3. *Show that the inverse of $L^{(i)}$ is obtained by simply multiplying the off-diagonal elements with -1. Show also that $(L^{(i)})^{-1}$ is an M-matrix, if A is an M-matrix.*

Exercise 13.4. *Prove that L (cf. (13.7)) is an M-matrix if A is an M-matrix.*

Exercise 13.5. *Show that if we ignore arbitrary off-diagonal elements in the $L^{(i)}$, the remaining lower triangular matrix L is still an M-matrix (if A is an M-matrix).*

We have in this way arrived at an incomplete LU decomposition of the M-matrix A. Obviously, the incomplete factors exist, but there is more. We have seen that ignoring Gaussian elimination elements in L and U has the effect that diagonal elements of the upper triangular factor U become larger and that the off-diagonal elements of the factors L and U become smaller in absolute value (but remain negative).

The following theorem collects the above results for the situation where we determine a priori the positions of the elements that we wish to ignore during the Gaussian elimination process. Note that this is not a serious restriction because we may also neglect elements during the process according to certain criteria and this defines the positions implicitly. The indices of the elements to be ignored are collected in a set $\mathcal{S}$:

$$S \subset S_n \equiv \{(i, j) | i \neq j, 1 \leq i \leq n, 1 \leq j \leq n\}. \tag{13.8}$$

We can now formulate the theorem that guarantees the existence of incomplete decompositions for the M-matrix A (cf. [139, Theorem 2.3]).

Theorem 13.3. *Let $A = (a_{i,j})$ be an $n \times n$ M-matrix, then there exists for every $S \subset S_n$ a lower triangular matrix $\widetilde{L} = (\ell_{i,j})$, with $\ell_{i,i} = 1$, an upper triangular matrix $\widetilde{U} = (u_{i,j})$, and a matrix $N = (n_{i,j})$ with*

- $\ell_{i,j} = 0$, $u_{i,j} = 0$, *if* $(i, j) \in S$
- $n_{i,j} = 0$ *if* $(i, j) \notin S$,

such that the splitting $A = \widetilde{L}\widetilde{U} - N$ leads to a convergent iteration (3.1). The factors $\widetilde{L}$ and $\widetilde{U}$ are uniquely defined by S.

We can, of course, make variations on these incomplete splittings, for instance, by isolating the diagonal of $\widetilde{U}$ as a separate factor. When A is symmetric and positive definite, then S is obviously selected so that it defines a symmetric sparsity pattern and then we can rewrite the factorization so that the diagonals of $\widetilde{L}$ and $\widetilde{U}$ are equal. These are known as incomplete Cholesky decompositions.

```
ILU for an n by n matrix A (cf. [10]):
for k = 1, 2, ..., n − 1
    d1/a_{k,k}
    for i = k + 1, k + 2, ..., n
      if (i, k) ∈ S
        e = d a_{i,k}; a_{i,k} = e
        for j = k + 1, ..., n
          if (i, j) ∈ S and (k, j) ∈ S
            a_{i,j} = a_{i,j} − e a_{k,j}
          end if
        end j
      end if
    end i
end k
```

Figure 13.1. ILU for a general matrix A.

A commonly used strategy is to define S by:

$$S = \{(i, j)| \quad a_{i,j} \neq 0\}. \tag{13.9}$$

That is, the only nonzeros allowed in the LU factors are those for which the corresponding entries in A are nonzero. It is easy to show that the elements $k_{i,j}$ of K match those of A on the set S:

$$k_{i,j} = a_{i,j} \quad \text{if} \quad (i, j) \in S. \tag{13.10}$$

Even though the conditions (13.1) and (13.10) together are sufficient (for certain classes of matrices) to determine the nonzero entries of L and U directly, it is more natural and simpler to compute these entries based on a simple modification of the Gaussian elimination algorithm; see Figure 13.1. The main difference from the usual Gaussian elimination algorithm is in the innermost j-loop where an update to $a_{i,j}$ is computed only if it is allowed by the constraint set S.

After the completion of the algorithm, the incomplete LU factors are stored in the corresponding lower and upper triangular parts of the array A. It can be shown that the computed LU factors satisfy (13.10).

The incomplete factors $\widetilde{L}$ and $\widetilde{U}$ define the preconditioner $K = (\widetilde{L}\widetilde{U})^{-1}$. In the context of an iterative solver, this means that we have to evaluate expressions like $z = (\widetilde{L}\widetilde{U})^{-1}y$ for any given vector y. This is done in two steps: first obtain w from the solution of $\widetilde{L}w = y$ and then compute z from $\widetilde{U}z = w$. Straightforward implementation of these processes leads to recursions, for which vector and parallel computers are not ideally suited. This sort of observation has led to reformulations of the preconditioner, for example, with reordering techniques or with blocking techniques. It has also led to different types of preconditioner, including diagonal scaling, polynomial preconditioning, and truncated Neumann series. These approaches may be useful in certain circumstances, but they tend to increase the computational complexity, because they often require more iteration steps or make each iteration step more expensive.

A well-known variant on ILU is the so-called *Modified ILU* (MILU) factorization [71, 105]. For this variant the condition (13.10) is replaced by

$$\sum_{j=1}^{n} k_{i,j} = \sum_{j=1}^{n} a_{i,j} + ch^2 \text{ for } i = 1, 2, \dots, n. \tag{13.11}$$

The term ch^2 is for grid-oriented problems with mesh-size h. Although in many applications this term is skipped (that is, we often take $c = 0$), this may lead to ineffective preconditioning [199, 195] or even breakdown of the preconditioner, see [72]. In our context, the row sum requirement in (13.11) amounts to an additional correction to the diagonal entries, compared to those computed in Figure 13.1 and to the d_i in (13.14). This correction leads to the observation that $Kz \approx Az$ for almost constant z (in fact this was the motivation for the construction of these preconditioners). This results in very fast convergence for problems where the solution is very smooth (almost constant for the components corresponding to the nonzero elements per row of A). However, quite the opposite may be observed for problems where the solution is far from smooth. For such problems MILU may lead to much slower convergence than ILU.

The incomplete factorizations have been generalized with blocks of A instead of single elements. The inverses of diagonal blocks in these incomplete block factorizations are themselves again approximated, for instance by their diagonal only or by the tridiagonal part, for details on this see [10, 144, 43]. In the author's experience, block incomplete decompositions can be quite effective for linear systems associated with 2-dimensional partial differential equations, discretized over rectangular grids. However, for 3-dimensional problems they appeared to be less effective.

13.2.1 An example of incomplete decompositions

I will illustrate the above sketched process for a popular type of preconditioner for sparse positive definite symmetric matrices, namely, the *Incomplete Cholesky factorization* [98, 139, 140, 211] with no fill-in. We will denote this preconditioner as IC(0). CG in combination with IC(0) is often referred to as ICCG(0). We shall consider IC(0) for the matrix with five nonzero diagonals that arises after the 5-point finite-difference discretization of the two-dimensional Poisson equation over a rectangular region, using a grid of dimensions n_x by n_y. If the entries of the three nonzero diagonals in the upper triangular part of A are stored in three arrays $a(\cdot, 1)$ for the main diagonal, $a(\cdot, 2)$ for the first co-diagonal, and $a(\cdot, 3)$ for the n_x-th co-diagonal, then the i-th row of the symmetric matrix A can be represented as in (13.12).

$$A = \begin{pmatrix} \ddots & & \ddots \quad \ddots \quad \ddots & & \ddots \\ & a_{i-n_x,3} & a_{i-1,2} \; a_{i,1} \; a_{i,2} & & a_{i,3} \\ & \ddots & \ddots \quad \ddots \quad \ddots & & \ddots \end{pmatrix} \tag{13.12}$$

Exercise 13.6. *If we write A as* $A = L_A + \mathrm{diag}(A) + L_A{}^T$, *in which* L_A *is the strictly lower triangular part of A, show that the IC(0)-preconditioner can be written as*

$$K = (L_A + D)D^{-1}(L_A{}^T + D). \tag{13.13}$$

Relation (13.13) only holds if there are no corrections to off-diagonal nonzero entries in the incomplete elimination process for A and if we ignore all fill-in outside the nonzero structure of A. It is easy to do this for the 5-point Laplacian. For other matrices, we can force the relation to hold only if we also ignore Gaussian elimination corrections at places where A has nonzero entries. This may decrease the effectiveness of the preconditioner, because we then neglect more operations in the Gaussian elimination process.

For IC(0), the entries d_i of the diagonal matrix D can be computed from the relation

$$\mathrm{diag}(K) = \mathrm{diag}(A).$$

For the 5-diagonal A, this leads to the following relations for the d_i:

$$d_i = a_{i,1} - a_{i-1,2}^2/d_{i-1} - a_{i-n_x,3}^2/d_{i-n_x}. \tag{13.14}$$

Obviously this is a recursion in both directions over the grid. This aspect will be discussed later when we consider the application of the preconditioner in the context of parallel and vector processing.

Axelsson and Lindskog [12] describe a relaxed form of modified incomplete decomposition that, for the 5-diagonal A, leads to the following relations for the d_i:

$$d_i = a_{i,1} - a_{i-1,2}(a_{i-1,2} + \alpha a_{i-1,3})/d_{i-1} \\ - a_{i-n_x,3}(a_{i-n_x,3} + \alpha a_{i-n_x,2})/d_{i-n_x}.$$

Note that, for $\alpha = 0$ we have the standard IC(0) decomposition, whereas for $\alpha = 1$ we have the modified Incomplete Cholesky decomposition MIC(0) proposed by Gustafsson [105]. It has been observed that, in many practical situations, $\alpha = 1$ does not lead to a reduction in the number of iteration steps, with respect to $\alpha = 0$, but in my experience, taking $\alpha = 0.95$ almost always reduces the number of iteration steps significantly [200]. The only difference between the IC(0) and MIC(0) is the choice of the diagonal D; in fact, the off-diagonal entries of the triangular factors are identical.

For the solution of systems $Kw = r$, given by

$$K^{-1}r = (L_A^T + D)^{-1} D (L_A + D)^{-1} r,$$

it will almost never be advantageous to determine the matrices $(L_A^T + D)^{-1}$ and $(L_A + D)^{-1}$ explicitly, since these matrices are usually dense triangular matrices. Instead, for the computation of, say, $y = (L_A + D)^{-1} r$, y is solved from the linear lower triangular system $(L_A + D)y = r$. This step then leads typically to relations for the entries y_i, of the form

$$y_i = (r_i - a_{i-1,2} y_{i-1} - a_{i-n_x,3} y_{i-n_x})/d_i,$$

which again represents a recursion in both directions over the grid, of the same form as the recursion for the d_i.

For differently structured matrices, we can also perform incomplete LU factorizations. Often, many of the ideas shown here for Incomplete Cholesky factorizations apply for efficient implementation. For more general matrices with the same nonzero structure as the 5-point Laplacian, some other well-known approximations lead to precisely the same type of recurrence relations as for Incomplete LU and Incomplete Cholesky: for example, Gauss–Seidel, SOR, SSOR [113], and SIP [183]. Hence these methods can often be made vectorizable or parallel in the same way as in the algorithm for Incomplete Cholesky preconditioning.

Since vector and parallel computers do not lend themselves well to recursions in a straightforward manner, the recursions just discussed may seriously degrade the effect of preconditioning on a vector or parallel computer, if carried out in the form given above. This sort of observation has led to different types of preconditioners, including diagonal scaling, polynomial preconditioning, and truncated Neumann series. Such approaches may be useful in certain circumstances, but they tend to increase the computational complexity (by requiring more iteration steps or by making each iteration step more expensive). On the other hand, various techniques have been proposed to vectorize the recursions, mainly based on reordering the unknowns or changing the order of computation. For regular grids, such approaches lead to highly vectorizable code for the standard incomplete factorizations (and consequently also for Gauss–Seidel, SOR, SSOR, and SIP). If our goal is to minimize computing time, there may thus be a trade-off between added complexity and increased vectorization. However, before discussing these techniques, I shall present a method of reducing the computational complexity of preconditioning.

13.2.2 Efficient implementations of ILU(0) preconditioning

Suppose that the given matrix A is written in the form $A = L_A + \mathrm{diag}(A) + U_A$, in which L_A and U_A are the strictly lower and upper triangular part of A, respectively. Eisenstat [74] has proposed an efficient implementation for preconditioned iterative methods, when the preconditioner K can be represented as

$$K = (L_A + D)D^{-1}(D + U_A), \tag{13.15}$$

in which D is a diagonal matrix. Some simple Incomplete Cholesky, incomplete LU, modified versions of these factorizations, as well as SSOR can be written in this form. For the incomplete factorizations, we have to ignore all the LU factorization corrections to off-diagonal entries [140]; the resulting decomposition is referred to as ILU(0) in the unsymmetric case, and IC(0) for the Incomplete Cholesky situation. For the 5-point finite-difference discretized operator over rectangular grids in two dimensions, this is equivalent to the incomplete factorizations with no fill-in, since in these situations there are no Gaussian elimination operations on nonzero off-diagonal entries.

The first step to make the preconditioning more efficient is to eliminate the diagonal D in (13.15). We rescale the original linear system $Ax = b$ to obtain

$$D^{-1/2}AD^{-1/2}\widetilde{x} = D^{-1/2}b, \tag{13.16}$$

or $\widetilde{A}\widetilde{x} = \widetilde{b}$, with $\widetilde{A} = D^{-1/2}AD^{-1/2}$, $\widetilde{x} = D^{1/2}x$, and $\widetilde{b} = D^{-1/2}b$. With $\widetilde{A} = L_{\widetilde{A}} + \mathrm{diag}(\widetilde{A}) + U_{\widetilde{A}}$, we can easily verify that

$$\widetilde{K} = (L_{\widetilde{A}} + I)(I + U_{\widetilde{A}}). \tag{13.17}$$

Note that the corresponding triangular systems, such as $(L_{\widetilde{A}} + I)r = w$, are more efficiently solved, since the division by the entries of D is avoided. Also note that this scaling does not necessarily have the effect that $\mathrm{diag}(\widetilde{A}) = I$.

The key idea in Eisenstat's approach (also referred to as *Eisenstat's trick*) is to apply standard iterative methods (that is, in their formulation with $K = I$) to the explicitly preconditioned linear system

$$(L_{\widetilde{A}} + I)^{-1}\widetilde{A}(I + U_{\widetilde{A}})^{-1}y = (L_{\widetilde{A}} + I)^{-1}\widetilde{b}, \tag{13.18}$$

where $y = (I + U_{\widetilde{A}})\widetilde{x}$. This explicitly preconditioned system will be denoted by $Py = c$. Now we can write $\widetilde{A}$ in the form

$$\widetilde{A} = L_{\widetilde{A}} + I + \mathrm{diag}(\widetilde{A}) - 2I + I + U_{\widetilde{A}}. \tag{13.19}$$

This expression, as well as the special form of the preconditioner given by (13.17), is used to compute the vector Pz for a given z by

$$Pz = (L_{\widetilde{A}} + I)^{-1}\widetilde{A}(I + U_{\widetilde{A}})^{-1}z = (L_{\widetilde{A}} + I)^{-1}(z + (\mathrm{diag}(\widetilde{A}) - 2I)t) + t, \tag{13.20}$$

with

$$t = (I + U_{\widetilde{A}})^{-1}z. \tag{13.21}$$

Note that the computation of Pz is equivalent to solving two triangular systems plus the multiplication of a vector by a diagonal matrix $(\mathrm{diag}(\widetilde{A}) - 2I)$ and an addition of this result to z. Therefore the matrix-vector product for the preconditioned system can be computed virtually at the cost of the matrix-vector product of the unpreconditioned system. This fact implies that the preconditioned system can be solved by any of the iterative methods for practically the same computational cost per iteration step as the unpreconditioned system. That is to say, the preconditioning comes essentially for free, in terms of computational complexity.

In most situations we see, unfortunately, that while we have avoided the fast part of the iteration process (the matrix-vector product Ap), we are left with the most problematic part of the computation, namely, the triangular solves. However, in some cases, as we shall see, these parts can also be optimized to about the same level of performance as the matrix-vector products.

13.3 Changing the order of computation

In some situations, it is possible to change the order of the computations without changing the results. A prime example is the ILU preconditioner for the

5-point finite-difference operator over a rectangular m by m grid. Suppose that we have indexed the unknowns according to their positions in the grid, lexicographically as $x_{1,1}, x_{1,2}, \ldots, x_{1,m}, x_{2,1}, \ldots, x_{m,m}$. Then, for the standard ILU(0) preconditioner, in which all fill-ins are discarded, it is easily verified that the computations for the unknowns $x_{i,j}$ can be done independently of each other along diagonals of the grid (grid points for which the sum of the indices is constant). This leads to vector code but, because there is only independence along each diagonal, the parallelism is too fine-grained. In three-dimensional problems, there are more possibilities to obtain vectorizable or parallel code. For the standard 7-point finite-difference approximation of elliptic PDEs over a regular rectangular grid, the equivalent of the diagonal in two dimensions is known as the hyperplane: a set of grid points for which the sum of the three indices is constant. It was reported in [171, 199] that this approach can lead to satisfactory performance on vector computers. For the CM-5 computer, a similar approach was developed in [27]. The obvious extension of hyperplanes (or diagonals) to irregular sparse matrices, defines the *wavefront ordering*, discussed in [161]. The success of a wavefront ordering depends very much on how well a given computer can handle indirect addressing. In general, the straightforward wavefront ordering approach gives too little opportunity for efficient parallelization.

Vuik, van Nooyen and Wesseling [214] generalize the wavefront approach to a block wavefront approach, using ideas that were originally proposed for parallel multigrid in [21]. They present results of experiments on a 128-processor CRAY 3TD. Van Duin [209, Chapter 3] uses graph concepts for the detection of parallelism. He attempts to identify strongly connected components for which independent ILU factorizations can be made. A drop tolerance strategy is used to create a large enough number of such components. This leads to the concept of MultiILU.

13.4 Reordering the unknowns

A standard trick for exploiting parallelism is to select all unknowns that have no direct relationship with each other and to number them first. For the 5-point finite-difference discretization over rectangular grids, this approach is known as a *red-black ordering*. For elliptic PDEs, this leads to very parallel preconditioners. The performance of the preconditioning step is as high as the performance of the matrix-vector product. However, changing the order of the unknowns leads in general to a different preconditioner. Duff and Meurant [69] report on experiments that show that most reordering schemes (for example, the red-black ordering) lead to a considerable increase in iteration steps (and hence in computing time) compared with the standard lexicographical ordering.

For the red-black ordering associated with the discretized Poisson equation, it can be shown that the condition number of the preconditioned system is only about one quarter that of the unpreconditioned system for ILU, MILU, and SSOR, with no asymptotic improvement as the gridsize h tends to zero [128].

One way to obtain a better balance between parallelism and fast convergence, is to use more colours [57]. In principle, since there is not necessarily any independence between different colours, using more colours decreases the parallelism but increases the global dependence and hence the convergence. In [58], up to 75 colours are used for a 76^2 grid on the NEC SX-3/14 resulting in a 2 Gflops performance, which is much better than for the wavefront ordering. With this large number of colours the speed of convergence for the preconditioned process is virtually the same as with a lexicographical ordering [57].

The concept of *multi-colouring* has been generalized to unstructured problems by Jones and Plassmann [123]. They propose effective heuristics for the identification of large independent subblocks of a given matrix. For problems large enough to get sufficient parallelism in these subblocks, their approach leads to impressive speedups compared to the natural ordering on a single processor.

Meier and Sameh [138] report on the parallelization of the preconditioned CG algorithm for a multivector processor with a hierarchical memory. Their approach is based on a red-black ordering in combination with forming a reduced system (Schur complement).

Another approach, suggested by Meurant [142], exploits the idea of the two-sided (or twisted) Gaussian elimination procedure for tridiagonal matrices. This is generalized for the incomplete factorization. Van der Vorst [198] has shown how this procedure can be done in a nested way. For 3D finite-difference problems, twisting can be used for each dimension, which gives an increase in parallelism by a factor of two per dimension. This leads, without further computational overheads, to an incomplete decomposition, as well as triangular solves, that can be done in eight parallel parts (2 in each dimension). For a discussion of these techniques see [61]. This parallel ordering technique is sometimes referred to as 'vdv' ordering [69], or 'van der Vorst' ordering, see for example [22].

Meurant [143] reports on timing results obtained on a CRAY Y-MP/832, using an incomplete repeated twisted block factorization for two-dimensional problems. For this approach for preconditioned CG, Meurant reports a speedup of nearly 6 on an 8-processor CRAY Y-MP. This speedup has been measured relative to the same repeated twisted factorization process executed on a single

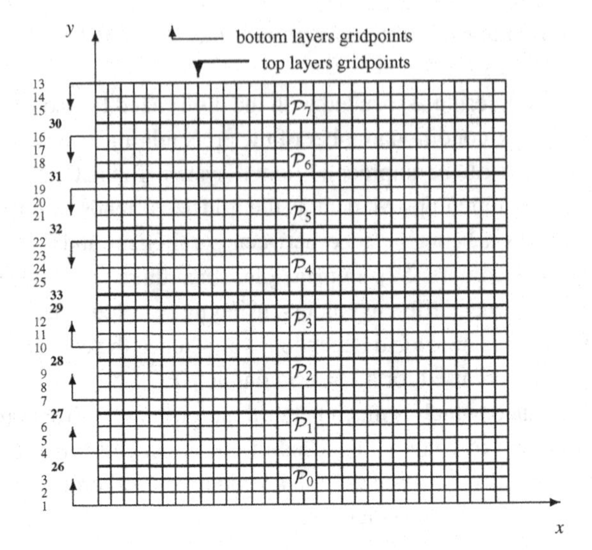

Figure 13.2. Decomposition of the grid into stripes, and assignment of subdomains to processors for $p = 8$. Arrows indicate the progressing direction of the line numbering per subdomain. Numbers along the y-axis give an example of global (line) ordering, which satisfies all the required conditions. Within each horizontal line, gridpoints are ordered lexicographically.

processor. Meurant also reports an increase in the number of iteration steps as a result of this repeated twisting. This increase implies that the effective speedup with respect to the best nonparallel code is only about 4.

A more sophisticated approach that combines ideas from twisting, domain decomposition with overlap, and reordering, was proposed in [132, 133, 134]. We will explain this idea for the special situation of a discretized second order elliptic PDE over a rectangular domain. The discretization has been carried out with the standard 5-point central difference stencil that leads, over a rectangular grid with lexicographical ordering, to the familiar block matrix with 5 nonzero diagonals.

The first step is to split the domain into blocks, as in domain decomposition methods, and to order the unknowns lexicographically per block. This has been indicated, for the case of 8 horizontal blocks, in Figure 13.2. Per block we start counting from one side ('the bottom layer'); the points on the last line ('the top layer') are ordered after all subdomains, as is indicated in Figure 13.3. For instance, the lines 1, 2, 3, and 26 all belong to the block stored with processor $\mathcal{P}_0$, but in the matrix interpretation the first 3 lines are ordered first and line 26 appears in the matrix only after all other 'interior' lines. This means that the matrix has the following nonzero structure (we give only a relevant part

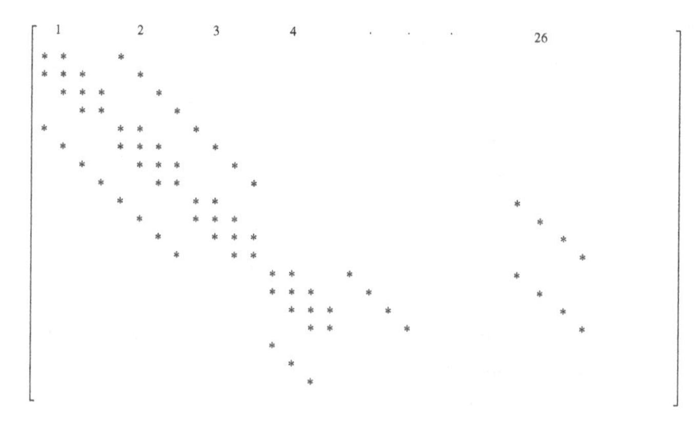

Figure 13.3. The structure of the reordered matrix.

of the matrix). Note that we have already introduced another element in our ordering, namely the idea of twisting: the lines of the subdomains are ordered from bottom to top and from top to bottom in Figure 13.2.

Now imagine what happens if we carry out an incomplete LU factorization with zero fill. That would create level-1 fill in the error matrix. Note that, in particular, we would introduce fill in the subblock of the matrix that connects line 26 with line 5, and note also that we would not have seen this level-1 fill if we had selected all points lexicographically.

Exercise 13.7. *Complete the structure in the matrix of Figure 13.3 and check where fill-in occurs due to incomplete pivoting. Identify the fill positions that lead to connections of unknowns that would not have occurred with lexicographical ordering.*

This means that if we want the block ordering to be at least as effective as the standard ordering, we have to remove this additional fill. This can be interpreted as permitting level-1 fill in a small overlap, and this is the reason for the name *pseudo-overlap* for this way of ordering. It is obvious how this idea is generalized for more arbitrary matrices: the new ordering is compared with the standard given one and the possibly additional level-1 fill is included in the preconditioner. The idea can also easily be applied to preconditioners with a higher level fill.

In [133, 132] an increase in the pseudo-overlap and also inclusion of higher levels of fill that are introduced by the new block-wise ordering are suggested. For high dimensional problems and relatively low numbers of processors this leads to almost negligible overhead. It is shown by analysis in [134] and by experiments [133, 132] that the block ordering with pseudo-overlap may lead to parallelizable incomplete decompositions that are almost perfectly scalable if the number of processors p is less than $\sqrt{(n)}$, where n denotes the order of the given linear system (the reported experiments include experiments for 16 processors, for $n \approx 260\,000$).

13.5 Variants of ILU preconditioners

Many variants on the idea of incomplete or modified incomplete decomposition have been proposed in the literature. These variants are designed to either reduce the total computational work or improve the performance on vector or parallel computers or handle special problems. We could, for instance, think of incomplete variants of the various LU-decomposition algorithms discussed in [98, Chapter 4.4].

I will describe some of the more popular variants and give references to where more details can be found for other variants.

A natural approach is to allow more fill-in in the LU factor (that is a larger set S), than those allowed by the condition (13.2). Several possibilities have been proposed. The most obvious variant is to allow more fill-ins in specific locations in the LU factors, for example allowing more nonzero bands in the $\widetilde{L}$ and $\widetilde{U}$ matrices (that is larger stencils) [11, 105, 140]. The most common location-based criterion is to allow a set number of levels of fill-in, where original entries have level zero, original zeros have level ∞ and a fill-in in position (i, j) has level determined by

$$Level_{ij} = \min_{1 \leq k \leq \min(i,j)} \{Level_{ik} + Level_{kj} + 1\}.$$

In the case of simple discretizations of partial differential equations, this gives a simple pattern for incomplete factorizations with different levels of fill-in. For example, if the matrix is from a 5-point discretization of the Laplacian in two dimensions, level 1 fill-in will give the original pattern plus a diagonal inside the outermost band (for instance, see [140]).

The other main criterion for deciding which entries to omit is to replace the *drop-by-position* strategy in (13.2) by a *drop-by-size* one. That is, a fill-in entry is discarded if its absolute value is below a certain threshold value. This *drop-tolerance* strategy was first proposed by [145]. For the regular problems just mentioned, it is interesting that the level fill-in and drop strategies give a somewhat similar incomplete factorization, because the numerical value of successive fill-in levels decreases markedly, reflecting the characteristic decay in the entries in the factors of the LU decomposition of A. For general problems, however, the two strategies can be significantly different. Since it is usually not known a priori how many entries will be above a selected threshold, the dropping strategy is normally combined with restricting the number of fill-ins allowed in each column [167]. When using a threshold criterion, it is possible to change it dynamically during the factorization to attempt to achieve a target density of the factors [13, 145]. Saad gives a very good overview of these techniques [168].

Although the notation is not yet fully standardized, the nomenclature commonly adopted for incomplete factorizations is ILU(k), when k levels of fill-in are allowed, and ILUT(α, f) for the threshold criterion, for which entries of modulus less than α are dropped and the maximum number of fill-ins allowed in any column is f. There are many variations on these strategies and the criteria are sometimes combined. In some cases, constraining the row sums of the incomplete factorization to match those of the matrix, as in MILU, can help [105], but as we noted earlier, successful application of this technique is restricted to cases where the solution of the (preconditioned) system is rather smooth.

Shifts can be introduced to prevent breakdown of the incomplete factorization process. As we have seen, incomplete decompositions exist for general M-matrices. It is well known that they may not exist if the matrix is positive definite, but does not have the M-matrix property.

Manteuffel [136] considered Incomplete Cholesky factorizations of diagonally shifted matrices. He proved that if A is symmetric positive definite, then there exists a constant $\alpha > 0$, such that the Incomplete Cholesky factorization of $A + \alpha I$ exists. Since we make an incomplete factorization for $A + \alpha I$, instead of A, it is not necessarily the case that this factorization is also efficient as a preconditioner; the only purpose of the shift is to avoid breakdown of the decomposition process. Whether there exist suitable values for α such that the preconditioner exists and is efficient is a matter of trial and error.

Another point of concern is that for non$M$-matrices the incomplete factors of A may be very ill-conditioned. For instance, it has been demonstrated in [194] that if A comes from a 5-point finite-difference discretization of $\Delta u + \beta(u_x + u_y) = f$, then for β sufficiently large, the incomplete LU factors may be very ill conditioned even though A has a very modest condition number. Remedies for reducing the condition numbers of $\widetilde{L}$ and $\widetilde{U}$ have been discussed in [76, 194].

13.6 Hybrid techniques

In the classical incomplete decompositions fill-in is ignored right from the start of the decomposition process. However, it might be a good idea to delay this until the matrix becomes too dense. This leads to a hybrid combination of direct and iterative techniques. One of such approaches has been described in [29]; I will describe it here in some detail.

We first permute the given matrix of the linear system $Ax = b$ to a doubly bordered block diagonal form:

$$\widetilde{A} = P^T A P = \begin{bmatrix} A_{00} & 0 & \cdots & 0 & A_{0m} \\ 0 & A_{11} & \ddots & \vdots & A_{1m} \\ \vdots & \ddots & \ddots & 0 & \vdots \\ 0 & \cdots & 0 & A_{m-1m-1} & \vdots \\ A_{m0} & A_{m1} & \cdots & \cdots & A_{mm} \end{bmatrix}. \tag{13.22}$$

Of course, the parallelism in the eventual method depends on the value of m, and some problems lend themselves more to this than others. Many circuit simulation problems can be rewritten in an effective way, as a circuit is often composed of components that are only locally coupled to others.

We also permute the right-hand side b to $\widetilde{b} = P^T b$, which leads to the system

$$\widetilde{A}\widetilde{x} = \widetilde{b}, \tag{13.23}$$

with $x = Px$.

The parts of $\widetilde{b}$ and $\widetilde{x}$ that correspond to the block ordering will be denoted by $\widetilde{b}_i$ and $\widetilde{x}_i$. The first step in the (parallelizable) algorithm will be to eliminate the unknown parts $\widetilde{x}_0, \cdots, \widetilde{x}_{m-1}$, which is done by the algorithm in Figure 13.4.

Note that S in Figure 13.4 denotes the Schur complement after the elimination of the blocks $0, 1, \ldots, m-1$. In many relevant situations, direct solution of the reduced system $Sx_m = y_m$ requires the dominating part of the total computational costs, and this is where we bring in the iterative component of the algorithm.

Exercise 13.8. *Suppose that we solve the reduced system $Sx_m = y_m$ with an iterative method and that after termination we have the approximated solution $\widehat{x}_m$, with $r_m = S\widehat{x}_m - y_m$. When we take this approximated solution for the computation of the x_i in Figure 13.4, then this leads to an approximated solution $\widehat{x}$ for the system $Ax = b$.*

Show that, in exact computation,

$$\|A\widehat{x} - b\|_2 = \|r_m\|_2. \tag{13.24}$$

Parallel_for $i = 0, 1, \ldots, m-1$
  Decompose A_{ii}: $A_{ii} = L_{ii}U_{ii}$
  $L_{mi} = A_{mi}U_{ii}^{-1}$
  $U_{im} = L_{ii}^{-1}A_{im}$
  $y_i = L_{ii}^{-1}\widetilde{b}_i$
  $S_i = L_{mi}U_{im}$
  $z_i = L_{mi}y_i$
end
$S = A_{mm} - \sum_{i=0}^{m-1} S_i$
$y_m = \widetilde{b}_m - \sum_{i=0}^{m-1} z_i$
Solve $Sx_m = y_m$
Parallel_for $i = 0, 1, \ldots, m-1$
  $x_i = U_{ii}^{-1}(y_i - U_{im}x_m)$
end

Figure 13.4. Parallel elimination.

The next step is to construct a preconditioner for the reduced system. This is based on discarding small elements in S. The elements larger than some threshold value define the preconditioner C:

$$c_{ij} = \begin{cases} s_{ij} & \text{if } |s_{ij} > t|s_{ii}| \text{ or } |s_{ij}| > t|s_{jj}| \\ 0 & elsewhere \end{cases} \tag{13.25}$$

with a parameter $0 \leq t < 1$. In the experiments reported in [29] the value $t = 0.02$ turned out to be satisfactory, but this may need some experimentation for specific problems.

When we take C as the preconditioner, then we have to solve systems like $Cv = w$, and this requires decomposition of C. In order to prevent too much fill-in, it is suggested that C is reordered with a minimum degree ordering. The system $Sx_m = y_m$ is then solved with, for instance, GMRES with preconditioner C. For the examples described in [29] it turns out that the convergence of GMRES was not very sensitive to the choice of t. The preconditioned iterative solution approach for the reduced system also offers opportunities for parallelism, although in [29] it is shown that even in serial mode the iterative solution (to sufficiently high precision) is often more efficient than direct solution of the reduced system.

Note that, because of (13.24), it is not necessary to iterate on the complete system.

In [29] heuristics are described for the decision on when the switch from direct to iterative should take place. These heuristics are based on mild assumptions on the speed of convergence of GMRES. The paper also reports on a number of experiments for linear systems, not only from circuit simulation, but also for some matrix problems taken from Matrix Market[2]. These experiments indicate that attractive savings in computational costs can be achieved, even in serial computation mode.

13.7 Element by element preconditioners

In finite-element problems, it is not always possible or sensible to assemble the entire matrix, and it is as easy to form products of the matrix with vectors as it is when held in assembled form. Furthermore, it is easy to distribute such matrix multiplications to exploit parallelism. Hence preconditioners are required that can be constructed at the element level. Hughes et al. [118] were the first to propose such *element by element* preconditioners.

A parallel variant is suggested in [106]. For symmetric positive definite A, they decompose each element matrix A_e as $A_e = L_e L_e^T$, and construct the preconditioner as $K = LL^T$, with

$$L = \sum_{e=1}^{n_e} L_e.$$

In this approach, nonadjacent elements can be treated in parallel. An overview and discussion of parallel element by element preconditioners is given in [210]. To our knowledge, the effectiveness of element by element preconditioners is limited, in the sense that often they do not give a substantial improvement of the CPU time.

13.8 Polynomial preconditioning

The main motivation for considering polynomial preconditioning is to improve the parallel performance of the solver, since the matrix-vector product is often more parallelizable than other parts of the solver (for instance the inner products). By doing so, all implementation tricks for the matrix-vector product can easily be exploited. The main problem is to find effective low degree polynomials $p_k(A)$, so that the iterative solver can be applied to $p_k(A)Ax = p_k(A)b$.

[2] Collection of test matrices available at
`ftp://ftp.cise.ufl.edu/cis/tech-reports/tr98/tr98-016.ps`.

With m steps of a Krylov solver, this leads to a Krylov subspace

$$K^m(p_k(A)A; r_0) = \text{span}(r_0, p_k(A)Ar_0, \ldots, (p_k(A)A)^{m-1}r_0),$$

and this is a subspace of the Krylov subspace $K^{(k+1)(m-1)+1}(A; r_0)$. The point is that we have arrived in a high dimensional subspace (with 'holes'), for the overhead costs of only m iteration steps. The hope is that for clever choices of p_k, this high dimensional subspace, with holes, will contain almost the same good approximation to the solution as the full Krylov subspace. If so, then we have saved ourselves all the overhead associated with the $(k + 1)(m - 1)$ iteration steps that are needed to create the full subspace.

A big problem with polynomial preconditioning is that the aforementioned 'holes' can cause us to miss important directions and so many more iterations are often required. Thus this form of preconditioning is usually only beneficial on a platform where inner products are expensive and for methods rich in inner products, like GMRES.

One approach for obtaining a polynomial preconditioner, reported in [65], is to use the low order terms of a Neumann expansion of $(I - B)^{-1}$, if A can be written as $A = I - B$ and the spectral radius of B is less than 1. It was suggested in [65] that a matrix splitting $A = K - N$ and a truncated power series for $K^{-1}N$ is used when the condition on B is not satisfied. More general polynomial preconditioners have also been proposed (see, for example, [7, 122, 163]). These polynomials are usually shifted Chebyshev polynomials over intervals that are estimated from the iteration parameters of a few steps of the unpreconditioned solver, or from other spectral information.

13.9 Sparse Approximate Inverse (SPAI)

The main reason why explicit inverses are not used is that, for irreducible matrices, the inverse will always be structurally dense. That is to say, sparse factorization techniques that produce sparse L and U will produce a dense inverse matrix even if most of the entries in the factors L and U are actually zero [66]. However, we may follow ideas similar to those used for the construction of ILU factorizations and compute and use directly a sparse approximation to the inverse. Perhaps the most obvious technique for this is to solve the problem[3]

$$\min_K ||I - AK||_F, \tag{13.26}$$

[3] We recall that $\| \, \|_F$ denotes the Frobenius norm of a matrix, viz. $\|A\|_F \equiv \sqrt{\sum_{i,j} a_{i,j}^2}$.

where K has some fully or partially prescribed sparsity structure. This problem can be expressed as n independent least-squares problems for each of the n columns of K. Each of these least-squares problems only involves a few variables and, because they are independent, they can be solved in parallel. With these techniques it is possible to successively increase the density of the approximation to reduce the value of (13.26) and so, in principle, ensure convergence of the preconditioned iterative method [46]. The small least-squares subproblems can be solved by standard (dense) QR factorizations [46, 99, 104].

In a further attempt to increase sparsity and reduce the computational cost of the solution of the subproblems, it has been suggested that a few steps of GMRES are used to solve the subsystems [39]. A recent study indicates that the computed approximate inverse may be a good alternative to ILU [99], but it is much more expensive to compute both in terms of time and storage, at least if computed sequentially. This means that it is normally only attractive to use this technique if the computational costs for the construction can be amortized by using the preconditioner for several right-hand sides. We have seen successful application of the SPAI approach for the solution of linear systems associated with electrical circuits, where different right-hand sides represent different voltage or current inputs.

One other problem with these approaches is that, although the residual for the approximation of a column of K can be controlled (albeit perhaps at the cost of a rather dense column in K), the nonsingularity of the matrix K is not guaranteed. Partly to avoid this, an approach that approximates the triangular factors of the inverse has been proposed [127]. The nonsingularity of the factors can be easily controlled and, if necessary, the sparsity pattern of the factors may also be controlled. Following this approach, sparse approximations to an A-biconjugate set of vectors using drop tolerances can be generated [24, 26]. In a scalar or vector environment, it is also much cheaper to generate the factors in this way than to solve the least-squares problems for the columns of the approximate inverse [25]. Van Duin [209, Chapter 5] shows how to compute (*sparsified*) inverses for incomplete Cholesky factors and Zhang [224] has developed a parallel preconditioning using incomplete triangular factors of the inverse.

Parallel implementation is almost an essential condition for efficient use of sparse approximate inverses. For publications that concentrate on that aspect see [18, 19, 23]. For highly structured matrices, some experiences have been reported in [103]. Gustafsson and Lindskog [107] have implemented a fully parallel preconditioner based on truncated Neumann expansions [196] to approximate the inverse SSOR factors of the matrix. Their experiments (on a CM-200) show a worthwhile improvement over a simple diagonal scaling.

Note that, because the inverse of the inverse of a sparse matrix is sparse, there are classes of dense matrices for which a sparse approximate inverse might be a very appropriate preconditioner. This may be the case for matrices that arise from boundary element type discretizations, for instance, in electromagnetism [1]. For some classes of problems, it may be attractive to construct the explicit inverses of the LU factors, even if these are considerably less sparse than the factors L and U, because such a factorization can be more efficient in parallel [3]. An incomplete form of this factorization for use as a preconditioner was proposed in [2].

For a good overview on the SPAI techniques see [22].

13.10 Preconditioning by blocks or domains

Other preconditioners that use direct methods, are those where the direct method, or an incomplete version of it, is used to solve a subproblem of the original problem. This can be done in *domain decomposition*, where problems on subdomains can be solved by a direct method but the interaction between the subproblems is handled by an iterative technique.

Domain decomposition methods were motivated by parallel computing, but it now appears that the approach can also be used with success for the construction of global preconditioners. This is usually done for linear systems that arise from the discretization of a PDE. The idea is to split the given domain into subdomains, and to compute an approximation for the solution on each subdomain. If all connections between subdomains are ignored, this then leads to a *Block Jacobi* preconditioner. Chan and Goovaerts [36] showed that the domain decomposition approach can actually lead to *improved* convergence rates, at least when the number of subdomains is not too large. This is because of the well-known divide and conquer effect when applied to methods with superlinear complexity such as ILU: it is more efficient to apply such methods to smaller problems and piece the global solution together.

In order to make the preconditioner more successful, the domains have to be coupled, that is we have to find proper boundary conditions along the interior boundaries of the subdomains. From a linear algebra point of view, this amounts to adapting the diagonal blocks in order to compensate for the neglected off-diagonal blocks. This is only successful if the matrix comes from a discretized partial differential equation and if certain smoothness conditions on the solution are assumed. If, for instance, the solution were constant, then we could remove the off-diagonal block entries adding them to the diagonal block entries without changing the solution. Likewise, if the solution is assumed to be fairly smooth along a domain interface, we might expect this technique of diagonal block correction to be effective. Domain decomposition is used in an iterative fashion and usually the interior boundary conditions (in matrix language: the corrections to diagonal blocks) are based upon information from the approximate solutions on the neighbouring subdomains that are available from a previous iteration step.

Tang [188] has proposed the concept of matrix enhancement. The idea is to introduce additional unknowns, which gives elegant possibilities for the formulation of effective domain decomposition of the underlying problem. For hyperbolic systems, this technique was further refined by Tan in [186] and by Tan and Borsboom [187]. I will explain the idea for the situation of two domains.

13.10.1 Canonical enhancement of a linear system

We start with the linear nonsingular system

$$\mathbf{B}\,\mathbf{y} = \mathbf{d}, \tag{13.27}$$

that results from discretization of a given PDE over some domain. Now, we partition the matrix $\mathbf{B}$, and the vectors $\mathbf{y}$ and $\mathbf{d}$ correspondingly,

$$\begin{bmatrix} \mathbf{B}_{11} & \mathbf{B}_{1\ell} & \mathbf{B}_{1r} & \mathbf{B}_{12} \\ \mathbf{B}_{\ell 1} & B_{\ell\ell} & B_{\ell r} & \mathbf{B}_{\ell 2} \\ \mathbf{B}_{r1} & B_{r\ell} & B_{rr} & \mathbf{B}_{r2} \\ \mathbf{B}_{21} & \mathbf{B}_{2\ell} & \mathbf{B}_{2r} & \mathbf{B}_{22} \end{bmatrix}, \quad \begin{bmatrix} \mathbf{y}_1 \\ y_\ell \\ y_r \\ \mathbf{y}_2 \end{bmatrix}, \quad \text{and} \quad \begin{bmatrix} \mathbf{d}_1 \\ d_\ell \\ d_r \\ \mathbf{d}_2 \end{bmatrix}. \tag{13.28}$$

The labels are not chosen arbitrarily: we associate with label 1 (and 2, respectively) elements/operations of the linear system corresponding to subdomain 1 (2, respectively) and with label ℓ (r, respectively) elements/operations corresponding to the left (right, respectively) of the interface between the two subdomains. The central blocks $B_{\ell\ell}$, $B_{\ell r}$, $B_{r\ell}$ and B_{rr} are square matrices of equal size, say, n_i by n_i. They correspond to the unknowns on the interface. Since the number of unknowns on the interface will typically be much smaller than the total number of unknowns, n_i will be much smaller than n, the size of $\mathbf{B}$.

For a typical discretization, the matrix $\mathbf{B}$ is banded and the unknowns are only locally coupled. Therefore it is not unreasonable to assume that $\mathbf{B}_{r1}$, $\mathbf{B}_{21}$, $\mathbf{B}_{12}$ and $\mathbf{B}_{\ell 2}$ are zero. For this situation, we define the 'canonical enhancement'

$\underline{\mathbf{B}}$ of $\mathbf{B}$, $\widetilde{\mathbf{y}}$ of $\mathbf{y}$, and $\underline{\mathbf{d}}$ of $\mathbf{d}$, by

$$\underline{\mathbf{B}} \equiv \left[\begin{array}{ccc|ccc} \mathbf{B}_{11} & \mathbf{B}_{1\ell} & \mathbf{B}_{1r} & \mathbf{0} & \mathbf{0} & \mathbf{0} \\ \mathbf{B}_{\ell 1} & B_{\ell\ell} & B_{\ell r} & 0 & 0 & \mathbf{0} \\ \mathbf{0} & I & 0 & -I & 0 & \mathbf{0} \\ \hline \mathbf{0} & 0 & -I & 0 & I & \mathbf{0} \\ \mathbf{0} & 0 & 0 & B_{r\ell} & B_{rr} & \mathbf{B}_{r2} \\ \mathbf{0} & \mathbf{0} & \mathbf{0} & \mathbf{B}_{2\ell} & \mathbf{B}_{2r} & \mathbf{B}_{22} \end{array}\right], \quad \widetilde{\mathbf{y}} \equiv \begin{bmatrix} \mathbf{y}_1 \\ y_\ell \\ \widetilde{y}_r \\ \widetilde{y}_\ell \\ y_r \\ \mathbf{y}_2 \end{bmatrix}, \tag{13.29}$$

and

$$\underline{\mathbf{d}} \equiv (\mathbf{d}_1, d_\ell, 0, 0, d_r, \mathbf{d}_2)^T .$$

We easily verify that $\mathbf{B}$ is also nonsingular and that $\underline{\mathbf{y}}$ is the unique solution of

$$\underline{\mathbf{B}}\,\widetilde{\mathbf{y}} = \underline{\mathbf{d}}, \tag{13.30}$$

with $\underline{\mathbf{y}} \equiv (\mathbf{y}_1^T, y_\ell^T, y_r^T, y_\ell^T, y_r^T, \mathbf{y}_2^T)^T$.

With this linear system we can associate a simple iterative scheme for the two coupled subblocks:

$$\begin{bmatrix} \mathbf{B}_{11} & \mathbf{B}_{1\ell} & \mathbf{B}_{1r} \\ \mathbf{B}_{\ell 1} & B_{\ell\ell} & B_{\ell r} \\ \mathbf{0} & I & 0 \end{bmatrix} \begin{bmatrix} \mathbf{y}_1^{(i+1)} \\ y_\ell^{(i+1)} \\ \widetilde{y}_r^{\,(i+1)} \end{bmatrix} = \begin{bmatrix} \mathbf{d}_1 \\ d_\ell \\ \widetilde{y}_\ell^{\,(i)} \end{bmatrix},$$

$$\begin{bmatrix} 0 & I & \mathbf{0} \\ B_{r\ell} & B_{rr} & \mathbf{B}_{r2} \\ \mathbf{B}_{2\ell} & \mathbf{B}_{2r} & \mathbf{B}_{22} \end{bmatrix} \begin{bmatrix} \widetilde{y}_\ell^{\,(i+1)} \\ y_r^{(i+1)} \\ \mathbf{y}_2^{\,(i+1)} \end{bmatrix} = \begin{bmatrix} \widetilde{y}_r^{\,(i)} \\ d_r \\ \mathbf{d}_2 \end{bmatrix}. \tag{13.31}$$

These systems can be solved in parallel and we recognize this as nothing else than a simple additive Schwartz iteration (with no overlap and Dirichlet–Dirichlet coupling). The extra unknowns $\widetilde{y}_\ell$ and $\widetilde{y}_r$, in the enhanced vector $\widetilde{\mathbf{y}}$, will serve for communication between the subdomains during the iterative solution process of the linear system. After termination of the iterative process, we have to undo the enhancement. We could simply skip the values of the additional elements, but since these also carry information one of the alternatives could be as follows.

With an approximate solution

$$\widetilde{\mathbf{y}}^{(i)} = (\mathbf{y}_1^{(i)^T}, y_\ell^{(i)^T}, (\widetilde{y}_r^{\,(i)})^T, (\widetilde{y}_\ell^{(i)})^T, y_r^{(i)^T}, \mathbf{y}_2^{(i)^T})^T$$

of (13.31), we may associate the approximate solution $\mathbf{R}\widetilde{\mathbf{y}}$ of (13.27) given by

$$\mathbf{R}\widetilde{\mathbf{y}} \equiv (\mathbf{y}_1^{(i)^T}, \tfrac{1}{2}(y_\ell^{(i)} + \widetilde{y}_\ell^{(i)})^T, \tfrac{1}{2}(y_r^{(i)} + \widetilde{y}_r^{(i)})^T, \mathbf{y}_2^{(i)^T})^T, \tag{13.32}$$

that is, we simply average the two sets of unknowns that should have been equal to each other at full convergence.

13.10.2 Interface coupling matrix

From (13.29) we see that the interface unknowns and the additional interface unknowns are coupled in a straightforward way by

$$\begin{bmatrix} I & 0 \\ 0 & -I \end{bmatrix} \begin{bmatrix} y_\ell \\ \widetilde{y}_r \end{bmatrix} = \begin{bmatrix} I & 0 \\ 0 & -I \end{bmatrix} \begin{bmatrix} \widetilde{y}_\ell \\ y_r \end{bmatrix}, \tag{13.33}$$

but, of course, we may replace the coupling matrix by any other nonsingular interface coupling matrix C:

$$C \equiv \begin{bmatrix} C_{\ell\ell} & -C_{\ell r} \\ -C_{r\ell} & C_{rr} \end{bmatrix}. \tag{13.34}$$

This leads to the following block system

$$\mathbf{B}_C\widetilde{\mathbf{y}} = \left[\begin{array}{ccc|ccc} \mathbf{B}_{11} & \mathbf{B}_{1\ell} & \mathbf{B}_{1r} & \mathbf{0} & \mathbf{0} & \mathbf{0} \\ \mathbf{B}_{\ell 1} & B_{\ell\ell} & B_{\ell r} & 0 & 0 & \mathbf{0} \\ \mathbf{0} & C_{\ell\ell} & C_{\ell r} & -C_{\ell\ell} & -C_{\ell r} & \mathbf{0} \\ \hline \mathbf{0} & -C_{r\ell} & -C_{rr} & C_{r\ell} & C_{rr} & \mathbf{0} \\ \mathbf{0} & 0 & 0 & B_{r\ell} & B_{rr} & \mathbf{B}_{r2} \\ \mathbf{0} & \mathbf{0} & \mathbf{0} & \mathbf{B}_{2\ell} & \mathbf{B}_{2r} & \mathbf{B}_{22} \end{array}\right] \begin{bmatrix} \mathbf{y}_1 \\ y_\ell \\ \widetilde{y}_r \\ \widetilde{y}_\ell \\ y_r \\ \mathbf{y}_2 \end{bmatrix} = \underline{\mathbf{d}}. \tag{13.35}$$

In a domain decomposition context, we will have for the approximate solution $\mathbf{y}$ that $\widetilde{y}_r \approx y_r$ and $\widetilde{y}_\ell \approx y_\ell$. If we know some analytic properties about the local behaviour of the true solution $\mathbf{y}$ across the interface, for instance, smoothness up to some degree, then we may try to identify a convenient coupling matrix C that takes advantage of this knowledge. We preferably want a C so that

$$-C_{\ell\ell}\widetilde{y}_\ell - C_{\ell r}y_r \approx -C_{\ell\ell}y_\ell - C_{\ell r}y_r \approx 0$$

$$\text{and} \quad - C_{r\ell}y_\ell - C_{rr}\widetilde{y}_r \approx -C_{r\ell}y_\ell - C_{rr}y_r \approx 0.$$

In that case (13.35) is almost decoupled into two independent smaller linear systems (identified by the two boxes).

The matrix consisting of the two blocks is then used as a preconditioner for the enhanced system. Tan [186, Chapters 2 and 3] (see also [187]) studied the interface conditions along boundaries of subdomains and forced continuity for the solution and some low-order derivatives at the interface. He also proposed including mixed derivatives in these relations, in addition to the conventional tangential and normal derivatives. The parameters involved are determined locally by means of normal mode analysis, and they are adapted to the discretized problem. It is shown that the resulting domain decomposition method defines a standard iterative method for some splitting $A = K - N$, and the local coupling aims to minimize the largest eigenvalues of $I - AK^{-1}$. Of course this method can be accelerated and impressive results for GMRES acceleration are shown in [186]. Some attention is paid to the case where the solutions for the subdomains are obtained with only modest accuracy per iteration step.

13.10.3 Other approaches

Washio and Hayami [217] employed a domain decomposition approach for a rectangular grid in which one step of SSOR is performed for the interior part of each subdomain. In order to make this domain-decoupled SSOR more like global SSOR, the SSOR iteration matrix for each subdomain is modified. In order to further improve the parallel performance, the inverses in these expressions are approximated by low-order truncated Neumann series. A similar approach is suggested in [217] for a block modified ILU preconditioner. Experimental results have been reported for a 32-processor NEC Cenju distributed memory computer.

Radicati and Robert [160] used an algebraic version of this approach by computing ILU factors within overlapping block diagonals of a given matrix A. When applying the preconditioner to a vector v, the values on the overlapped region are taken as the average of the two values computed by the overlapping ILU factors. The approach of Radicati and Robert has been further refined by de Sturler [53], who studies the effects of overlap from the point of view of geometric domain decomposition. He introduces artificial mixed boundary conditions on the internal boundaries of the subdomains. In [53] (Table 5.8), experimental results are shown for a decomposition into 20×20 slightly overlapping subdomains of a 200×400 mesh for a discretized convection-diffusion equation (5-point stencil). Using a twisted ILU preconditioning on each subdomain, it is shown that the complete linear system can be solved by GMRES on a 400-processor distributed memory Parsytec system with an efficiency of about 80% (this means that, with this domain adapted preconditioner, the process is about 320 times faster than ILU preconditioned GMRES for the unpartitioned linear system on a single processor). Since twisting leads to more parallelism, we can use bigger blocks (which usually means a better approximation). This helps to explain the good results.

Haase [110] suggests constructing an incomplete Cholesky decomposition on each subdomain and modifying the decomposition using information from neighbouring subdomains. His results, for the discretized Poisson equation in 3D, show that an increase in the number of domains scarcely affects the effectiveness of the preconditioner. Experimental results for a realistic finite-element model, on a 16-processor Parsytec Xplorer, show very good scalability of the Conjugate Gradient method with this preconditioner.

Heisse and Jung [114] attempt to improve the effectiveness of a domain decomposition preconditioner by using a multigrid V-cycle with only one pre- and one post-smoothing step of a parallel variant of Gauss–Seidel type to solve a coarse grid approximation to the problem. With the usual domain decomposition technique, effects of local changes in a domain that lead to global changes in the solution travel a distance of only one layer of neighbouring domains per iteration. The coarse grid corrections are used to pass this globally relevant information more quickly to all domains. The combination with Conjugate Gradients, which is the underlying method used for the local subproblems, leads to good results on a variety of platforms, including a 64-processor GC Power Plus machine. In fact, they use combinations of multigrid methods and iterative methods.

For general systems, we could apply a block Jacobi preconditioning to the normal equations, which would result in the block Cimmino algorithm [4]. A similar relationship exists between a block SOR preconditioning and the block Kaczmarz algorithm [30]. Block preconditioning for symmetric systems is discussed in [43]; in [45] incomplete factorizations are used within the diagonal blocks. Attempts have been made to preorder matrices to put large entries into the diagonal blocks so that the inverse of the matrix is well approximated by the block diagonal matrix whose block entries are the inverses of the diagonal blocks [38]. In fact, it is possible to have a significant effect on the convergence of these methods just by permuting the matrix to put large entries on the diagonal and then scaling it to reduce the magnitude of off-diagonal entries [68].

References

[1] G. Alléon, M. Benzi, and L. Giraud. Sparse approximate inverse preconditioning for dense linear systems arising in computational electromagnetics. *Numerical Algorithms*, 16(1):1–15, 1997.

[2] F. L. Alvarado and H. Dağ. Incomplete partitioned inverse preconditioners. Technical report, Department of Electrical and Computer Engineering, University of Wisconsin, Madison, 1994.

[3] F. L. Alvarado and R. Schreiber. Optimal parallel solution of sparse triangular systems. *SIAM J. Sci. Comput.*, 14:446–460, 1993.

[4] M. Arioli, I. S. Duff, J. Noailles, and D. Ruiz. A block projection method for sparse equations. *SIAM J. Sci. Statist. Comput.*, 13:47–70, 1992.

[5] M. Arioli, I. S. Duff, and D. Ruiz. Stopping criteria for iterative solvers. *SIAM J. Matrix Anal. Applic.*, 13:138–144, 1992.

[6] W. E. Arnoldi. The principle of minimized iteration in the solution of the matrix eigenproblem. *Quart. Appl. Math.*, 9:17–29, 1951.

[7] S. F. Ashby. Minimax polynomial preconditioning for Hermitian linear systems. *SIAM J. Matrix Anal. Applic.*, 12:766–789, 1991.

[8] O. Axelsson. Solution of linear systems of equations: iterative methods. In V. A. Barker, editor, *Sparse Matrix Techniques*, pages 1–51, Berlin, 1977. Springer-Verlag.

[9] O. Axelsson. Conjugate gradient type methods for unsymmetric and inconsistent systems of equations. *Linear Alg. Appl.*, 29:1–16, 1980.

[10] O. Axelsson. *Iterative Solution Methods.* Cambridge University Press, Cambridge, 1994.

[11] O. Axelsson and V.A. Barker. *Finite Element Solution of Boundary Value Problems. Theory and Computation.* Academic Press, New York, NY, 1984.

[12] O. Axelsson and G. Lindskog. On the eigenvalue distribution of a class of preconditioning methods. *Numer. Math.*, 48:479–498, 1986.

[13] O. Axelsson and N. Munksgaard. Analysis of incomplete factorizations with fixed storage allocation. In D. Evans, editor, *Preconditioning Methods – Theory and Applications*, pages 265–293. Gordon and Breach, New York, 1983.

[14] O. Axelsson and P. S. Vassilevski. A black box generalized conjugate gradient solver with inner iterations and variable-step preconditioning. *SIAM J. Matrix Anal. Applic.*, 12(4):625–644, 1991.

[15] Z. Bai and J. W. Demmel. Using the matrix *sign* function to compute invariant subspaces. *SIAM J. Matrix Anal. Applic.*, 1998:205–225, 1998.

[16] Z. Bai, D. Hu, and L. Reichel. A Newton basis GMRES implementation. *IMA J. Numer. Anal.*, 14:563–581, 1991.

[17] R. E. Bank and T. F. Chan. An analysis of the composite step biconjugate gradient method. *Numer. Math.*, 66:295–319, 1993.

[18] S. T. Barnard, L. M. Bernardo, and H. D. Simon. An MPI implementation of the SPAI preconditioner on the T3E. Technical Report LBNL-40794 UC405, Lawrence Berkeley National Laboratory, 1997.

[19] S. T. Barnard and R. L. Clay. A portable MPI implementation of the SPAI preconditioner in ISIS++. In M. Heath et al., editor, *Proceedings of the Eighth SIAM Conference on Parallel Processing for Scientific Computing*. SIAM, 1997.

[20] R. Barrett, M. Berry, T. Chan, J. Demmel, J. Donato, J. Dongarra, V. Eijkhout, R. Pozo, C. Romine, and H. van der Vorst. *Templates for the Solution of Linear Systems: Building Blocks for Iterative Methods*. SIAM, Philadelphia, PA, 1994.

[21] P. Bastian and G. Horton. Parallelization of robust multigrid methods: ILU factorization and frequency decomposition method. *SIAM J. Sci. Statist. Comput.*, 6:1457–1470, 1991.

[22] M. Benzi. Preconditioning techniques for large linear systems: A survey. *J. Comput. Phys.*, 2003. To appear.

[23] M. Benzi, J. Marín, and M. Tůma. A two-level parallel preconditioner based on sparse approximate inverses. In D. R. Kincaid and A. C. Elster, editors, *Iterative Methods in Scientific Computing, II*, pages 167–178. IMACS, 1999.

[24] M. Benzi, C. D. Meyer, and M. Tůma. A sparse approximate inverse preconditioner for the conjugate gradient method. *SIAM J. Sci. Comput.*, 17:1135–1149, 1996.

[25] M. Benzi and M. Tůma. Numerical experiments with two sparse approximate inverse preconditioners. *BIT*, 38:234–241, 1998.

[26] M. Benzi and M. Tůma. A sparse approximate inverse preconditioner for nonsymmetric linear systems. *SIAM J. Sci. Comput.*, 19(3):968–994, 1998.

[27] H. Berryman, J. Saltz, W. Gropp, and R. Mirchandaney. Krylov methods preconditioned with incompletely factored matrices on the CM-2. *J. Par. Dist. Comp.*, 8:186–190, 1990.

[28] A. Björck and T. Elfving. Accelerated projection methods for computing pseudoinverse solutions of systems of linear equations. *BIT*, 19:145–163, 1979.

[29] C. W. Bomhof and H. A. van der Vorst. A parallel linear system solver for circuit-simulation problems. *Num. Lin. Alg. Appl.*, 7:649–665, 2000.

[30] R. Bramley and A. Sameh. Row projection methods for large nonsymmetric linear systems. *SIAM J. Sci. Statist. Comput.*, 13:168–193, 1992.

[31] C. Brezinski. *Projection Methods for Systems of Equations*. North-Holland, Amsterdam, 1997.

[32] C. Brezinski and M. Redivo Zaglia. Treatment of near breakdown in the CGS algorithm. *Numerical Algorithms*, 7:33–73, 1994.

[33] A. M. Bruaset. *A Survey of Preconditioned Iterative Methods*. Longman Scientific and Technical, Harlow, UK, 1995.

[34] G. Brussino and V. Sonnad. A comparison of direct and preconditioned iterative techniques for sparse unsymmetric systems of linear equations. *Int. J. for Num. Methods in Eng.*, 28:801–815, 1989.

[35] B. L. Buzbee, G. H. Golub, and C. W. Nielson. On direct methods for solving Poisson's equations. *SIAM J. Numer. Anal.*, 7:627–656, 1970.

[36] T. F. Chan and D. Goovaerts. A note on the efficiency of domain decomposed incomplete factorizations. *SIAM J. Sci. Statist. Comput.*, 11:794–803, 1990.

[37] T. F. Chan and H. A. van der Vorst. Approximate and incomplete factorizations. In D. E. Keyes, A. Sameh, and V. Venkatakrishnan, editors, *Parallel Numerical Algorithms*, ICASE/LaRC Interdisciplinary Series in Science and Engineering, pages 167–202. Kluwer, Dordrecht, 1997.

[38] H. Choi and D. B. Szyld. Threshold ordering for preconditioning nonsymmetric problems with highly varying coefficients. Technical Report 96-51, Department of Mathematics, Temple University, Philadelphia, 1996.

[39] E. Chow and Y. Saad. Approximate inverse preconditioners via sparse-sparse iterations. *SIAM J. Sci. Comput.*, 19:995–1023, 1998.

[40] A. T. Chronopoulos and C. W. Gear. s-Step iterative methods for symmetric linear systems. *J. Comp. and Appl. Math.*, 25:153–168, 1989.

[41] A. T. Chronopoulos and S. K. Kim. s-Step Orthomin and GMRES implemented on parallel computers. Technical Report 90/43R, UMSI, Minneapolis, 1990.

[42] P. Concus and G. H. Golub. A generalized Conjugate Gradient method for nonsymmetric systems of linear equations. Technical Report STAN-CS-76-535, Stanford University, Stanford, CA, 1976.

[43] P. Concus, G. H. Golub, and G. Meurant. Block preconditioning for the conjugate gradient method. *SIAM J. Sci. Statist. Comput.*, 6:220–252, 1985.

[44] P. Concus, G. H. Golub, and D. P. O'Leary. A generalized conjugate gradient method for the numerical solution of elliptic partial differential equations. In J. R. Bunch and D. J. Rose, editors, *Sparse Matrix Computations*. Academic Press, New York, 1976.

[45] P. Concus and G. Meurant. On computing INV block preconditionings for the conjugate gradient method. *BIT*, pages 493–504, 1986.

[46] J. D. F. Cosgrove, J. C. Diaz, and A. Griewank. Approximate inverse preconditionings for sparse linear systems. *Intern. J. Computer Math.*, 44:91–110, 1992.

[47] G. C. (Lianne) Crone. The conjugate gradient method on the Parsytec GCel-3/512. *FGCS*, 11:161–166, 1995.

[48] L. Crone and H. van der Vorst. Communication aspects of the conjugate gradient method on distributed-memory machines. *Supercomputer*, X(6):4–9, 1993.

[49] J. Cullum and A. Greenbaum. Relations between Galerkin and norm-minimizing iterative methods for solving linear systems. *SIAM J. Matrix Anal. Applic.*, 17:223–247, 1996.

[50] J. W. Daniel, W. B. Gragg, L. Kaufmann, and G. W. Stewart. Reorthogonalization and stable algorithms for updating the Gram–Schmidt QR factorization. *Math. Comp.*, 30:772–795, 1976.

[51] E. de Sturler. A parallel restructured version of GMRES(m). Technical Report 91-85, Delft University of Technology, Delft, 1991.

[52] E. de Sturler. A parallel variant of GMRES(m). In R. Miller, editor, *Proc. of the Fifth Int. Symp. on Numer. Methods in Eng.*, 1991.

[53] E. de Sturler. *Iterative methods on distributed memory computers*. PhD thesis, Delft University of Technology, Delft, the Netherlands, 1994.

[54] E. de Sturler and D. R. Fokkema. Nested Krylov methods and preserving the orthogonality. In N. Duane Melson, T. A. Manteuffel, and S. F. McCormick, editors, *Sixth Copper Mountain Conference on Multigrid Methods*, NASA Conference Publication 3324, Part I, pages 111–126. NASA, 1993.

[55] E. de Sturler and H.A. van der Vorst. Reducing the effect of global communication in GMRES(m) and CG on parallel distributed memory computers. *Applied Numerical Mathematics*, 18:441–459, 1995.

[56] J. Demmel, M. Heath, and H. van der Vorst. Parallel numerical linear algebra. In *Acta Numerica 1993*, pages 111–197. Cambridge University Press, Cambridge, 1993.

[57] S. Doi. On parallelism and convergence of incomplete LU factorizations. *Applied Numerical Mathematics*, 7:417–436, 1991.

[58] S. Doi and A. Hoshi. Large numbered multicolor MILU preconditioning on SX-3/14. *Int'l. J. Computer Math.*, 44:143–152, 1992.

[59] J. J. Dongarra. Performance of various computers using standard linear equations software in a Fortran environment. Technical Report CS-89-85, University of Tennessee, Knoxville, 1990.

[60] J. J. Dongarra, I. S. Duff, D. C. Sorensen, and H. A. van der Vorst. *Solving Linear Systems on Vector and Shared Memory Computers*. SIAM, Philadelphia, PA, 1991.

[61] J. J. Dongarra, I. S. Duff, D. C. Sorensen, and H. A. van der Vorst. *Numerical Linear Algebra for High-Performance Computers*. SIAM, Philadelphia, PA, 1998.

[62] J. J. Dongarra, V. Eijkhout, and H. A. van der Vorst. Iterative solver benchmark. Technical report, University of Tennessee, 2000.

[63] Jack J. Dongarra and Henk A. van der Vorst. Performance of various computers using standard sparse linear equations solving techniques. *Supercomputer*, 9(5):17–29, 1992.

[64] J. Drkosova, A. Greenbaum, M. Rozložník, and Z. Strakoš. Numerical stability of GMRES. *BIT*, 35:309–330, 1995.

[65] P. F. Dubois, A. Greenbaum, and G. H. Rodrigue. Approximating the inverse of a matrix for use in iterative algorithms on vector processors. *Computing*, 22:257–268, 1979.

[66] I. S. Duff, A. M. Erisman, C. W. Gear, and J. K. Reid. Sparsity structure and Gaussian elimination. *SIGNUM Newsletter*, 23(2):2–8, 1988.

[67] I. S. Duff, A. M. Erisman, and J. K. Reid. *Direct methods for sparse matrices*. Oxford University Press, London, 1986.

[68] I. S. Duff and J. Koster. The design and use of algorithms for permuting large entries to the diagonal of sparse matrices. *SIAM J. Matrix Anal. Applic.*, 20:889–901, 1999.

[69] I. S. Duff and G. A. Meurant. The effect of ordering on preconditioned conjugate gradient. *BIT*, 29:635–657, 1989.

[70] I. S. Duff and H. A. van der Vorst. Developments and trends in the parallel solution of linear systems. *Parallel Computing*, 25:1931–1970, 1999.

[71] T. Dupont, R. P. Kendall, and H. H. Rachford Jr. An approximate factorization procedure for solving self-adjoint elliptic difference equations. *SIAM J. Numer. Anal.*, 5(3):559–573, 1968.

[72] V. Eijkhout. Beware of unperturbed modified incomplete point factorizations. In R. Beauwens and P. de Groen, editors, *Iterative Methods in Linear Algebra*, pages 583–591, Amsterdam, 1992. North-Holland.

[73] T. Eirola and O. Nevanlinna. Accelerating with rank-one updates. *Linear Alg. Appl.*, 121:511–520, 1989.

[74] S. C. Eisenstat. Efficient implementation of a class of preconditioned conjugate gradient methods. *SIAM J. Sci. Statist. Comput.*, 2(1):1–4, 1981.

[75] H. C. Elman. *Iterative methods for large sparse nonsymmetric systems of linear equations*. PhD thesis, Yale University, New Haven, CT, 1982.

[76] H. C. Elman. Relaxed and stabilized incomplete factorizations for non-self-adjoint linear systems. *BIT*, 29:890–915, 1989.

[77] M. Embree. *Convergence of Krylov subspace methods for non-normal matrices*. PhD thesis, Oxford University Computing Laboratory, Oxford, 1999.

[78] M. Engeli, T. Ginsburg, H. Rutishauser, and E. Stiefel. *Refined Iterative Methods for Computation of the Solution and the Eigenvalues of Self-Adjoint Boundary Value Problems*. Birkhäuser, Basel/Stuttgart, 1959.

[79] V. Faber and T. A. Manteuffel. Necessary and sufficient conditions for the existence of a conjugate gradient method. *SIAM J. Numer. Anal.*, 21(2):352–362, 1984.

[80] K. Fan. Note on M-matrices. *Quart. J. Math. Oxford Ser. (2)*, 11:43–49, 1960.

[81] B. Fischer. *Orthogonal Polynomials and Polynomial Based Iteration Methods for Indefinite Linear Systems*. PhD thesis, University of Hamburg, Hamburg, Germany, 1994.

[82] B. Fischer. *Polynomial based iteration methods for symmetric linear systems*. Advances in Numerical Mathematics. Wiley and Teubner, Chichester, Stuttgart, 1996.

[83] R. Fletcher. *Conjugate gradient methods for indefinite systems*, volume 506 of *Lecture Notes Math.*, pages 73–89. Springer-Verlag, Berlin–Heidelberg–New York, 1976.

[84] D. R. Fokkema, G. L. G. Sleijpen, and H. A. van der Vorst. Generalized conjugate gradient squared. *J. Comp. and Appl. Math.*, 71:125–146, 1996.

[85] R. W. Freund. Conjugate gradient-type methods for linear systems with complex symmetric coefficient matrices. *SIAM J. Sci. Comput.*, 13:425–448, 1992.

[86] R. W. Freund. Quasi-kernel polynomials and their use in non-Hermitian matrix iterations. *J. Comp. and Appl. Math.*, 43:135–158, 1992.

[87] R. W. Freund. A transpose-free quasi-minimal residual algorithm for non-Hermitian linear systems. *SIAM J. Sci. Comput.*, 14:470–482, 1993.

[88] R. W. Freund, G. H. Golub, and N. M. Nachtigal. Iterative solution of linear systems. In *Acta Numerica 1992*, pages 57–100. Cambridge University Press, Cambridge, 1992.

[89] R. W. Freund, M. H. Gutknecht, and N. M. Nachtigal. An implementation of the look-ahead Lanczos algorithm for non-Hermitian matrices. *SIAM J. Sci. Comput.*, 14:137–158, 1993.

[90] R. W. Freund and N. M. Nachtigal. An implementation of the look-ahead Lanczos algorithm for non-Hermitian matrices, part 2. Technical Report 90.46, RIACS, NASA Ames Research Center, 1990.

[91] R. W. Freund and N. M. Nachtigal. QMR: a quasi-minimal residual method for non-Hermitian linear systems. *Numer. Math.*, 60:315–339, 1991.

[92] E. Gallopoulos and Y. Saad. Efficient solution of parabolic equations by Krylov approximation methods. *SIAM J. Sci. Statist. Comput.*, 13:1236–1264, 1992.

[93] C. F. Gauss. *Werke, Band IX*. Teubner, Leipzig, 1903.

[94] T. Ginsburg. Contribution I/5: The Conjugate Gradient method. In J. H. Wilkinson and C. Reinsch, editors, *Handbook of Automatic Computation, Linear Algebra, Vol II*. Springer Verlag, Berlin, 1971.

[95] G. H. Golub and W. Kahan. Calculating the singular values and pseudo-inverse of a matrix. *SIAM J. Numer. Anal.*, 2:205–224, 1965.

[96] G. H. Golub and D. P. O'Leary. Some history of the conjugate gradient and Lanczos algorithms: 1948–1976. *SIAM Rev.*, 31:50–102, 1989.

[97] G. H. Golub and H. A. van der Vorst. Numerical progress in eigenvalue computation in the 20th century. *J. Comp. and Appl. Math.*, 123(1-2):35–65, 2000.

[98] G. H. Golub and C. F. Van Loan. *Matrix Computations*. The Johns Hopkins University Press, Baltimore, 1996.

[99] N. I. M. Gould and J. A. Scott. Sparse approximate-inverse preconditioners using norm-minimization techniques. *SIAM J. Sci. Comput.*, 19(2):605–625, 1998.

[100] A. Greenbaum. Estimating the attainable accuracy of recursively computed residual methods. *SIAM J. Matrix Anal. Applic.*, 18:535–551, 1997.

[101] A. Greenbaum. *Iterative Methods for Solving Linear Systems*. SIAM, Philadelphia, 1997.

[102] A. Greenbaum, V. Ptak, and Z. Strakoš. Any nonincreasing convergence curve is possible for GMRES. *SIAM J. Matrix Anal. Applic.*, 17:465–469, 1996.

[103] M. Grote and H. Simon. Parallel preconditioning and approximate inverses on the connection machine. In R. F. Sincovec, D. E. Keyes, M. R. Leuze, L. R. Petzold, and D. A. Reed, editors, *Proceedings of the Sixth SIAM Conference on Parallel Processing for Scientific Computing*, pages 519–523, Philadelphia, 1993. SIAM.

[104] M. J. Grote and T. Huckle. Parallel preconditionings with sparse approximate inverses. *SIAM J. Sci. Comput.*, 18:838–853, 1997.

[105] I. Gustafsson. A class of first order factorization methods. *BIT*, 18:142–156, 1978.

[106] I. Gustafsson and G. Lindskog. A preconditioning technique based on element matrix factorizations. *Comput. Methods Appl. Mech. Eng.*, 55:201–220, 1986.

[107] I. Gustafsson and G. Lindskog. Completely parallelizable preconditioning methods. *Num. Lin. Alg. Appl.*, 2:447–465, 1995.

[108] M. H. Gutknecht. A completed theory of the unsymmetric Lanczos process and related algorithms, Part I. *SIAM J. Matrix Anal. Applic.*, 13:594–639, 1992.

[109] M. H. Gutknecht. Variants of BiCGStab for matrices with complex spectrum. *SIAM J. Sci. Comput.*, 14:1020–1033, 1993.

[110] G. Haase. Parallel incomplete Cholesky preconditioners based on the nonoverlapping data distribution. *Parallel Computing*, 24:1685–1703, 1998.

[111] W. Hackbusch. *Iterative Lösung großer schwachbesetzter Gleichungssysteme*. Teubner, Stuttgart, 1991.

[112] W. Hackbusch. *Elliptic Differential Equations*. Springer-Verlag, Berlin, 1992.

[113] L. A. Hageman and D. M. Young. *Applied Iterative Methods*. Academic Press, New York, 1981.

[114] B. Heisse and M. Jung. Parallel solvers for nonlinear elliptic problems based on domain decomposition ideas. *Parallel Computing*, 22:1527–1544, 1997.

[115] M. R. Hestenes and E. Stiefel. Methods of conjugate gradients for solving linear systems. *J. Res. Natl. Bur. Stand.*, 49:409–436, 1952.

[116] N. J. Higham. *Accuracy and Stability of Numerical Algorithms*. SIAM, Philadelphia, PA, 1996.

[117] M. Hochbruck and C. Lubich. On Krylov subspace approximations to the matrix exponential operator. *SIAM J. Numer. Anal.*, 34:1911–1925, 1997.

[118] T. J. R. Hughes, I. Levit, and J. Winget. An element-by-element solution algorithm for problems of structural and solid mechanics. *J. Comp. Methods in Appl. Mech. Eng.*, 36:241–254, 1983.

[119] C. P. Jackson and P. C. Robinson. A numerical study of various algorithms related to the preconditioned Conjugate Gradient method. *Int. J. for Num. Meth. in Eng.*, 21:1315–1338, 1985.

[120] D. A. H. Jacobs. Preconditioned Conjugate Gradient methods for solving systems of algebraic equations. Technical Report RD/L/N 193/80, Central Electricity Research Laboratories, 1981.

[121] K. C. Jea and D. M. Young. Generalized conjugate-gradient acceleration of nonsymmetrizable iterative methods. *Linear Alg. Appl.*, 34:159–194, 1980.

[122] O. G. Johnson, C. A. Micheli, and G. Paul. Polynomial preconditioning for conjugate gradient calculations. *SIAM J. Numer. Anal.*, 20:363–376, 1983.

[123] M. T. Jones and P. E. Plassmann. The efficient parallel iterative solution of large sparse linear systems. In A. George, J. R. Gilbert, and J. W. H. Liu, editors, *Graph Theory and Sparse Matrix Computations*, IMA Vol 56. Springer-Verlag, Berlin, 1994.

[124] E. F. Kaasschieter. A practical termination criterion for the Conjugate Gradient method. *BIT*, 28:308–322, 1988.

[125] E. F. Kaasschieter. Preconditioned conjugate gradients for solving singular systems. *J. Comp. and Appl. Math.*, 24:265–275, 1988.

[126] C. T. Kelley. *Iterative Methods for Linear and Nonlinear Equations*. SIAM, Philadelphia, PA, 1995.

[127] L. Yu. Kolotilina and A. Yu. Yeremin. Factorized sparse approximate inverse preconditionings. *SIAM J. Matrix Anal. Applic.*, 14:45–58, 1993.

[128] J. C. C. Kuo and T. F. Chan. Two-color Fourier analysis of iterative algorithms for elliptic problems with red/black ordering. *SIAM J. Sci. Statist. Comput.*, 11:767–793, 1990.

[129] C. Lanczos. An iteration method for the solution of the eigenvalue problem of linear differential and integral operators. *J. Res. Natl. Bur. Stand.*, 45:225–280, 1950.

[130] C. Lanczos. Solution of systems of linear equations by minimized iterations. *J. Res. Natl. Bur. Stand.*, 49:33–53, 1952.

[131] J. Liesen, M. Rozložník, and Z. Strakoš. Least squares residuals and minimal residual methods. *SIAM J. Sci. Comput.*, 23:1503–1525, 2002.

[132] M. Magolu monga Made and H. A. van der Vorst. A generalized domain decomposition paradigm for parallel incomplete LU factorization preconditionings. *Future Generation Computer Systems*, 17:925–932, 2001.

[133] M. Magolu monga Made and H. A. van der Vorst. Parallel incomplete factorizations with pseudo-overlapped subdomains. *Parallel Computing*, 27:989–1008, 2001.

[134] M. Magolu monga Made and H. A. van der Vorst. Spectral analysis of parallel incomplete factorizations with implicit pseudo-overlap. *Num. Lin. Alg. Appl.*, 9:45–64, 2002.

[135] T. A. Manteuffel. The Tchebychev iteration for nonsymmetric linear systems. *Numer. Math.*, 28:307–327, 1977.

[136] T. A. Manteuffel. An incomplete factorization technique for positive definite linear systems. *Math. Comp.*, 31:473–497, 1980.

[137] K. Meerbergen and M. Sadkane. Using Krylov approximations to the matrix exponential operator in Davidson's method. *Applied Numerical Mathematics*, 31:331–351, 1999.

[138] U. Meier and A. Sameh. The behavior of conjugate gradient algorithms on a multivector processor with a hierarchical memory. Technical Report CSRD 758, University of Illinois, Urbana, IL, 1988.

[139] J. A. Meijerink and H. A. van der Vorst. An iterative solution method for linear systems of which the coefficient matrix is a symmetric M-matrix. *Math. Comp.*, 31:148–162, 1977.

[140] J. A. Meijerink and H. A. van der Vorst. Guidelines for the usage of incomplete decompositions in solving sets of linear equations as they occur in practical problems. *J. Comput. Phys.*, 44:134–155, 1981.

[141] G. Meurant. The block preconditioned conjugate gradient method on vector computers. *BIT*, 24:623–633, 1984.

[142] G. Meurant. Numerical experiments for the preconditioned conjugate gradient method on the CRAY X-MP/2. Technical Report LBL-18023, University of California, Berkeley, CA, 1984.

[143] G. Meurant. The conjugate gradient method on vector and parallel supercomputers. Technical Report CTAC-89, University of Brisbane, July 1989.

[144] G. Meurant. *Computer Solution of Large Linear Systems*. North-Holland, Amsterdam, 1999.

[145] N. Munksgaard. Solving sparse symmetric sets of linear equations by preconditioned conjugate gradient method. *ACM Trans. Math. Softw.*, 6:206–219, 1980.

[146] N. M. Nachtigal, S. C. Reddy, and L. N. Trefethen. How fast are nonsymmetric matrix iterations? *SIAM J. Matrix Anal. Applic.*, 13:778–795, 1992.

[147] H. Neuberger. The overlap Dirac operator. In B. Medeke, A. Frommer, Th. Lippert and K. Schilling, editors, *Numerical Challenges in Lattice Quantum*

Chromodynamics, pages 1–17, Berlin, 2000. Springer-Verlag. Lecture notes in Computational Science and Engineering, 15.

[148] J. M. Ortega. *Introduction to Parallel and Vector Solution of Linear Systems*. Plenum Press, New York and London, 1988.

[149] C. C. Paige. Computational variants of the Lanczos method for the eigenproblem. *J. Inst. Math. Appl.*, 10:373–381, 1972.

[150] C. C. Paige. Error analysis of the Lanczos algorithm for tridiagonalizing a symmetric matrix. *J. Inst. Math. Appl.*, 18:341–349, 1976.

[151] C. C. Paige. Accuracy and effectiveness of the Lanczos algorithm for the symmetric eigenproblem. *Linear Alg. Appl.*, 34:235–258, 1980.

[152] C. C. Paige, B. N. Parlett, and H. A. van der Vorst. Approximate solutions and eigenvalue bounds from Krylov subspaces. *Num. Lin. Alg. Appl.*, 2(2):115–134, 1995.

[153] C. C. Paige and M. A. Saunders. Solution of sparse indefinite systems of linear equations. *SIAM J. Numer. Anal.*, 12:617–629, 1975.

[154] C. C. Paige and M. A. Saunders. LSQR: An algorithm for sparse linear equations and sparse least squares. *ACM Trans. Math. Softw.*, 8:43–71, 1982.

[155] B. N. Parlett. *The Symmetric Eigenvalue Problem*. Prentice-Hall, Englewood Cliffs, NJ, 1980.

[156] B. N. Parlett, D. R. Taylor, and Z. A. Liu. A look-ahead Lanczos algorithm for unsymmetric matrices. *Math. Comp.*, 44:105–124, 1985.

[157] C. Pommerell. *Solution of large unsymmetric systems of linear equations*. PhD thesis, Swiss Federal Institute of Technology, Zürich, 1992.

[158] C. Pommerell and W. Fichtner. PILS: An iterative linear solver package for ill-conditioned systems. In *Supercomputing '91*, pages 588–599, Los Alamitos, CA., 1991. IEEE Computer Society.

[159] A. Quarteroni and A. Valli. *Numerical Approximation of Partial Differential Equations*. Springer-Verlag, Berlin, 1994.

[160] G. Radicati di Brozolo and Y. Robert. Parallel conjugate gradient-like algorithms for solving sparse non-symmetric systems on a vector multiprocessor. *Parallel Computing*, 11:223–239, 1989.

[161] G. Radicati di Brozolo and M. Vitaletti. Sparse matrix-vector product and storage representations on the IBM 3090 with Vector Facility. Technical Report 513-4098, IBM-ECSEC, Rome, July 1986.

[162] J. Roberts. Linear model reduction and solution of algebraic Riccati equations. *Inter. J. Control*, 32:677–687, 1980.

[163] Y. Saad. Practical use of polynomial preconditionings for the conjugate gradient method. *SIAM J. Sci. Statist. Comput.*, 6:865–881, 1985.

[164] Y. Saad. Krylov subspace methods on supercomputers. Technical report, RIACS, Moffett Field, CA, September 1988.

[165] Y. Saad. *Numerical methods for large eigenvalue problems*. Manchester University Press, Manchester, UK, 1992.

[166] Y. Saad. A flexible inner-outer preconditioned GMRES algorithm. *SIAM J. Sci. Comput.*, 14:461–469, 1993.

[167] Y. Saad. ILUT: A dual threshold incomplete LU factorization. *Num. Lin. Alg. Appl.*, 1:387–402, 1994.

[168] Y. Saad. *Iterative Methods for Sparse Linear Systems.* PWS Publishing Company, Boston, 1996.

[169] Y. Saad and M. H. Schultz. GMRES: a generalized minimal residual algorithm for solving nonsymmetric linear systems. *SIAM J. Sci. Statist. Comput.*, 7:856–869, 1986.

[170] Y. Saad and H. A. van der Vorst. Iterative solution of linear systems in the 20-th century. *J. Comp. and Appl. Math.*, 123 (1-2):1–33, 2000.

[171] J. J. F. M. Schlichting and H. A. van der Vorst. Solving 3D block bidiagonal linear systems on vector computers. *J. Comp. and Appl. Math.*, 27:323–330, 1989.

[172] A. Sidi. Efficient implementation of minimal polynomial and reduced rank extrapolation methods. *J. Comp. and Appl. Math.*, 36:305–337, 1991.

[173] H. D. Simon. Direct sparse matrix methods. In James C. Almond and David M. Young, editors, *Modern Numerical Algorithms for Supercomputers*, pages 325–444, Austin, 1989. The University of Texas at Austin, Center for High Performance Computing.

[174] G. L. G. Sleijpen and D. R. Fokkema. BiCGSTAB(ℓ) for linear equations involving unsymmetric matrices with complex spectrum. *ETNA*, 1:11–32, 1993.

[175] G. L. G. Sleijpen and H. A. van der Vorst. Maintaining convergence properties of BICGSTAB methods in finite precision arithmetic. *Numerical Algorithms*, 10:203–223, 1995.

[176] G. L. G. Sleijpen and H. A. van der Vorst. Reliable updated residuals in hybrid Bi-CG methods. *Computing*, 56:141–163, 1996.

[177] G. L. G. Sleijpen, H. A. van der Vorst, and D. R. Fokkema. Bi-CGSTAB(ℓ) and other hybrid Bi-CG methods. *Numerical Algorithms*, 7:75–109, 1994.

[178] G. L. G. Sleijpen, H. A. van der Vorst, and J. Modersitzki. The main effects of rounding errors in Krylov solvers for symmetric linear systems. Technical Report Preprint 1006, Utrecht University, Department of Mathematics, 1997.

[179] G. L. G. Sleijpen, H. A. van der Vorst, and J. Modersitzki. Differences in the effects of rounding errors in Krylov solvers for symmetric indefinite linear systems. *SIAM J. Matrix Anal. Applic.*, 22(3):726–751, 2000.

[180] P. Sonneveld. CGS: a fast Lanczos-type solver for nonsymmetric linear systems. *SIAM J. Sci. Statist. Comput.*, 10:36–52, 1989.

[181] G. W. Stewart. *Matrix Algorithms, Vol. I: Basic Decompositions.* SIAM, Philadelphia, 1998.

[182] G. W. Stewart. *Matrix Algorithms, Vol. II: Eigensystems.* SIAM, Philadelphia, 2001.

[183] H. S. Stone. Iterative solution of implicit approximations of multidimensional partial differential equations. *SIAM J. Numer. Anal.*, 5:530–558, 1968.

[184] R. Sweet. A parallel and vector variant of the cyclic reduction algorithm. *Supercomputer*, 22:18–25, 1987.

[185] D. B. Szyld and J. A. Vogel. FQMR: A flexible quasi-minimal residual method with inexact preconditioning. *SIAM J. Sci. Comput.*, 23:363–380, 2001.

[186] K. H. Tan. *Local coupling in domain decomposition.* PhD thesis, Utrecht University, Utrecht, the Netherlands, 1995.

[187] K. H. Tan and M. J. A. Borsboom. On generalized Schwarz coupling applied to advection-dominated problems. In *Domain decomposition methods in*

scientific and engineering computing, pages 125–130, Providence, RI, 1993. Amer. Math. Soc.

[188] W. P. Tang. Generalized Schwarz splitting. *SIAM J. Sci. Statist. Comput.*, 13:573–595, 1992.

[189] C. H. Tong and Q. Ye. Analysis of the finite precision Bi-Conjugate Gradient algorithm for nonsymmetric linear systems. *Math. Comp.*, 69:1559–1575, 2000.

[190] J. van den Eshof, A. Frommer, Th. Lippert, K. Schilling, and H. A. van der Vorst. Numerical methods for the QCD overlap operator: I. sign-function and error bounds. *Computer Physics Communications*, 146:203–224, 2002.

[191] A. van der Sluis and H. A. van der Vorst. The rate of convergence of conjugate gradients. *Numer. Math.*, 48:543–560, 1986.

[192] A. van der Sluis and H. A. van der Vorst. Numerical solution of large sparse linear algebraic systems arising from tomographic problems. In G. Nolet, editor, *Seismic Tomography*, chapter 3, pages 49–83. Reidel Pub. Comp., Dordrecht, 1987.

[193] A. van der Sluis and H. A. van der Vorst. SIRT- and CG-type methods for the iterative solution of sparse linear least-squares problems. *Linear Alg. Appl.*, 130:257–302, 1990.

[194] H. A. van der Vorst. Iterative solution methods for certain sparse linear systems with a non-symmetric matrix arising from PDE-problems. *J. Comp. Phys.*, 44:1–19, 1981.

[195] H. A. van der Vorst. *Preconditioning by Incomplete Decompositions*. PhD thesis, Utrecht University, Utrecht, the Netherlands, 1982.

[196] H. A. van der Vorst. A vectorizable variant of some ICCG methods. *SIAM J. Sci. Statist. Comput.*, 3:86–92, 1982.

[197] H. A. van der Vorst. An iterative solution method for solving $f(A)x = b$, using Krylov subspace information obtained for the symmetric positive definite matrix a. *J. Comp. and Appl. Math.*, 18:249–263, 1987.

[198] H. A. van der Vorst. Large tridiagonal and block tridiagonal linear systems on vector and parallel computers. *Parallel Computing*, 5:45–54, 1987.

[199] H. A. van der Vorst. High performance preconditioning. *SIAM J. Sci. Statist. Comput.*, 10:1174–1185, 1989.

[200] H. A. van der Vorst. ICCG and related methods for 3D problems on vector computers. *Computer Physics Communications*, 53:223–235, 1989.

[201] H. A. van der Vorst. Bi-CGSTAB: A fast and smoothly converging variant of Bi-CG for the solution of non-symmetric linear systems. *SIAM J. Sci. Statist. Comput.*, 13:631–644, 1992.

[202] H. A. van der Vorst. Conjugate gradient type methods for nonsymmetric linear systems. In R. Beauwens and P. de Groen, editors, *Iterative Methods in Linear Algebra*, pages 67–76, Amsterdam, 1992. North-Holland.

[203] H. A. van der Vorst. Computational methods for large eigenvalue problems. In P. G. Ciarlet and J. L. Lions, editors, *Handbook of Numerical Analysis*, volume VIII, pages 3–179. North-Holland, Amsterdam, 2002.

[204] H. A. van der Vorst and J. B. M. Melissen. A Petrov–Galerkin type method for solving $ax = b$, where a is symmetric complex. *IEEE Trans. on Magn.*, pages 706–708, 1990.

[205] H. A. van der Vorst and G. L. G. Sleijpen. The effect of incomplete decomposition preconditioning on the convergence of Conjugate Gradients. In W. Hackbusch and G. Wittum, editors, *Incomplete Decompositions (ILU) – Algorithms, Theory, and Applications*, pages 179–187, Braunschweig, 1993. Vieweg. vol 41.

[206] H. A. van der Vorst and C. Vuik. The superlinear convergence behaviour of GMRES. *J. Comp. and Appl. Math.*, 48:327–341, 1993.

[207] H. A. van der Vorst and C. Vuik. GMRESR: A family of nested GMRES methods. *Num. Lin. Alg. Appl.*, 1:369–386, 1994.

[208] H. A. van der Vorst and Q. Ye. Refined residual replacement techniques for subspace iterative methods for convergence of true residuals. *SIAM J. Sci. Comput.*, 22:835–852, 2000.

[209] A. C. N. van Duin. *Parallel Sparse Matrix Computations*. PhD thesis, Utrecht University, Utrecht, the Netherlands, 1998.

[210] M. B. van Gijzen. *Iterative solution methods for linear equations in finite element computations*. PhD thesis, Delft University of Technology, Delft, the Netherlands, 1994.

[211] R. S. Varga. Factorizations and normalized iterative methods. In R. E. Langer, editor, *Boundary Problems in Differential Equations*, pages 121–142, Madison, WI, 1960. University of Wisconsin Press.

[212] R. S. Varga. *Matrix Iterative Analysis*. Prentice-Hall, Englewood Cliffs NJ, 1962.

[213] P. K. W. Vinsome. ORTHOMIN: an iterative method for solving sparse sets of simultaneous linear equations. In *Proc. Fourth Symposium on Reservoir Simulation*, pages 149–159, 1976.

[214] C. Vuik, R. R. P. van Nooyen, and P. Wesseling. Parallelism in ILU-preconditioned GMRES. *Parallel Computing*, 24:1927–1946, 1998.

[215] H. F. Walker. Implementation of the GMRES method using Householder transformations. *SIAM J. Sci. Statist. Comput.*, 9:152–163, 1988.

[216] H. F. Walker and L. Zhou. A simpler GMRES. *Num. Lin. Alg. Appl.*, 1:571–581, 1994.

[217] T. Washio and K. Hayami. Parallel block preconditioning based on SSOR and MILU. *Num. Lin. Alg. Appl.*, 1:533–553, 1994.

[218] R. Weiss. Error-minimizing Krylov subspace methods. *SIAM J. Sci. Comput.*, 15:511–527, 1994.

[219] R. Weiss. *Parameter-Free Iterative Linear Solvers*. Akademie Verlag, Berlin, 1996.

[220] P. Wesseling. *An Introduction to Multigrid Methods*. John Wiley and Sons, Chichester, 1992.

[221] O. Widlund. A Lanczos method for a class of nonsymmetric systems of linear equations. *SIAM J. Numer. Anal.*, 15:801–812, 1978.

[222] J. H. Wilkinson. *The Algebraic Eigenvalue Problem*. Clarendon Press, Oxford, 1965.

[223] J. H. Wilkinson and C. Reinsch, editors. *Handbook for Automatic Computation, Vol 2, Linear Algebra*. Springer-Verlag, Heidelberg – Berlin – New York, 1971.

[224] J. Zhang. A sparse approximate inverse technique for parallel preconditioning of general sparse matrices. Technical Report 281-98, Department of Computer Science, University of Kentucky, KY, 1998.

[225] Shao-Liang Zhang. GPBi-CG: Generalized product-type methods based on Bi-CG for solving nonsymmetric linear systems. *SIAM J. Sci. Comput.*, 18:537–551, 1997.

[226] L. Zhou and H. F. Walker. Residual smoothing techniques for iterative methods. *SIAM J. Sci. Statist. Comput.*, 15:297–312, 1994.

Index

Printed in the United States
by Baker & Taylor Publisher Services